Praise for *The Tao of Vegetable Gardening*

"If you want to read the complete, deepest-down lowdown on how to grow ~~~~
vegetables successfully, this is the book. It also stands as a
ine, independent lifestyle possible, relying only on nature ar
detailed knowledge of plant life to achieve successful food pr
way of life. The reader learns not only how to grow and ~
to breed new varieties and save the seed. And while you re~
charmed with the Tao philosophy of living—something I have come to believe is a
sure path to tranquility."

GENE LOGSDON, author, *Gene Everlasting* and *The Contrary Farmer*

"*The Tao of Vegetable Gardening* is another absolutely brilliant book from Carol Deppe.
It's smart, ultimately sensible, refreshing in the way old assumptions get questioned,
vastly informative about gardening—plus it's a really good read. I mean, how many
gardening books make you laugh out loud and get you to pick up the phone and order
a tool from a place called Red Pig? I'm so grateful for this book—I will have it memo-
rized by the time the soil is ready to work."

DEBORAH MADISON, author, *Vegetable Literacy*

"There are many knowledgeable gardeners but very few wise ones. Carol Deppe is both.
Her excellent new book, *The Tao of Vegetable Gardening,* serves up generous portions
of homegrown know-how gleaned from three decades' worth of experimentation. It
will, no doubt, make you a better gardener. What sets this book apart, though, is its
potential for making us into *happier* gardeners by sharing the deeper life lessons our
gardens have to teach. The Chinese word *tao* can be defined in different ways but my
favorite is "path," and Carol Deppe shows us that the timeless path to health, happi-
ness, and wholeness cuts right through our own backyard, if we choose to take it."

ROGER DOIRON, founding director, Kitchen Gardeners International

"Why do different ripe tomatoes harvested from the same plant in the same season
taste different? What does bean seed color have to do with vigor and flavor? After
nearly forty years in the seed business, I still learn amazing things from each new
book by Carol Deppe. *The Tao of Vegetable Gardening* melds the observational skills
and curiosity of a molecular geneticist with the sheer joy and inner harmonies of a
practicing participant in the garden's dance of life."

CR LAWN, founder, Fedco Seeds

THE TAO OF VEGETABLE GARDENING

THE TAO OF VEGETABLE GARDENING

*Cultivating Tomatoes,
Greens, Peas, Beans, Squash,
Joy, and Serenity*

CAROL DEPPE

CHELSEA GREEN PUBLISHING
WHITE RIVER JUNCTION, VERMONT

Unless otherwise noted, all photographs copyright © by Carol Deppe.

All *Tao Te Ching* passages are from *Tao Te Ching: A Window to the Tao through the Words of Lao Tzu*, copyright © 2010 by Carol Deppe. All the Taoist stories are from *Taoist Stories: A Window to the Tao through the Tales of Chuang Tzu and Lieh Tzu*, copyright © 2013 by Carol Deppe.

Project Manager: Bill Bokermann
Developmental Editor: Benjamin Watson
Copy Editor: Laura Jorstad
Proofreader: Michelle Moran
Indexer: Barbara Mortenson

Printed in The United States of America.
First printing January, 2015.
10 9 8 7 6 5 4 3 2 1   15 16 17 18

Our Commitment to Green Publishing

Chelsea Green sees publishing as a tool for cultural change and ecological stewardship. We strive to align our book manufacturing practices with our editorial mission and to reduce the impact of our business enterprise in the environment. We print our books and catalogs on chlorine-free recycled paper, using vegetable-based inks whenever possible. This book may cost slightly more because it was printed on paper that contains recycled fiber, and we hope you'll agree that it's worth it. Chelsea Green is a member of the Green Press Initiative (www.greenpressinitiative.org), a nonprofit coalition of publishers, manufacturers, and authors working to protect the world's endangered forests and conserve natural resources. *The Tao of Vegetable Gardening* was printed on paper supplied by Thomson-Shore that contains 100% postconsumer recycled fiber.

Library of Congress Cataloging-in-Publication Data
Deppe, Carol, author.
The Tao of vegetable gardening : cultivating tomatoes, greens, peas,
beans, squash, joy, and serenity / Carol Deppe.
     pages cm
Includes index.
ISBN 978-1-60358-487-6 (pbk.) -- ISBN 978-1-60358-488-3 (ebook) 1.
Vegetable gardening. I. Title.

SB321.D43 2015
635--dc23

2014036464

Chelsea Green Publishing
85 North Main Street, Suite 120
White River Junction, VT 05001
(802) 295-6300
www.chelseagreen.com

in memory of
Christina ("Kit") Ward

beloved editor, literary agent,
mentor, and friend

CONTENTS

ACKNOWLEDGMENTS

I thank Ben Watson for his expert, gentle editing of this, our third gardening book together.

I thank Margo Baldwin, my very hands-on and innovative publisher, and the rest of the Chelsea Green Publishing team for contributing their myriad skills, experience, wisdom, and wizardry.

For information, encouragement, company, and camaraderie I thank Charlotte Anthony, Nancy Baumeister, Nick Estens, Nate France, Paul Harcombe, Nonie Harcombe, Bruce Hecht, Mike Hessel, Alan Kapuler, Linda Kapuler, CR Lawn, Rose Marie Nichols McGee, Keane McGee, Stewart Pollack, Dane Rogers, and Merry Youle.

For believing in me and supporting my work I thank Denise-Christine, Mark Deppe, Kathy Ging, Kinsey Green, Sarah Kleeger, James Rodell, Nick Routledge, Janet Russell, Jerry Russell, Andrew Still, Roger Trevisiol, Helen Trevisiol, and Janice Wilson.

Anything I accomplish in this world is in its turn part of the accomplishments of my mentors, whom I pause to remember and acknowledge here: my science mentors, geneticist and orchid breeder Henry Wallbrunn, biochemist Arthur L. Koch, and fungal geneticist John R. Raper; my writing mentors Roger Swain and Christina ("Kit") Ward.

This book is dedicated to Kit Ward, one of the great editor-mentors and later, agent-mentors. Kit passed away in fall of 2012. She is sorely missed. But she lives on in her family and friends, and in the authors she helped shape, taught, nurtured, and guided.

Finally, I thank Colleen Mohyde of Doe Coover Agency for being there after Kit's death with caring, comfort, and encouragement.

INTRODUCTION

The word *tao* includes the concepts of way, path, method, subject, art, science, force, Spirit, God, power, and essence. *The Tao of Vegetable Gardening* is about the practical methods as well as the deeper essence of gardening. In this book I focus on growing food, especially tomatoes, green beans, peas, and leafy greens, and I illustrate the principles and practice of gardening primarily in the context of these vegetables. This allows me to discuss these crops in more depth than is common in gardening books that cover a little bit of everything. In my previous book, *The Resilient Gardener: Food Production and Self-Reliance in Uncertain Times*, I covered the major staple garden and homestead crops—potatoes, corn, dry beans, squash, and eggs. That book also has major chapters on managing soil fertility and watering; those topics are not repeated here.

For the beginning gardener I cover how to start, grow, and plant transplants in the tomato chapter. The same methods apply to all other transplants. I cover how to direct-seed big and small seeds in the chapters on peas and beans and on greens, respectively. With the methods in those chapters plus the general chapters in this book you can plant and grow just about everything.

In this book I introduce the eat-all greens garden approach to raising greens. I discovered the approach almost by accident and have been exploring and developing it further for more than two decades. It is by far the easiest, most space-saving and labor-efficient way of growing greens. With this method a family can raise all their summer greens as well as freeze and dry enough greens for winter with even a tiny garden. I believe the eat-all greens garden has the potential to completely transform the growing of greens in small or urban gardens. It also has commercial potential. I envision a frozen food industry of the future that offers a dozen different delicious frozen greens at prices considerably lower than anything currently possible.

With the eat-all greens garden approach I also finally achieve what has long been my dream—crops and methods that allow a gardener to do nothing whatsoever after sowing the seed until it is time to come back and harvest.

Almost every gardener loves and grows tomatoes, and there is nothing that beats the spectacular variety and intensity of flavor of heirloom tomatoes. However, all our current much-beloved heirloom tomato varieties are now threatened by the spread of new and more virulent lines of late blight, the same disease that was the scourge of Ireland in the Great Potato Famine. In parts of Europe it has already become impossible to grow tomatoes outdoors unless they have serious genetic blight resistance; likewise it is becoming difficult in some years to grow tomatoes in some gardens on the East Coast of the United States. It is likely that, within the next decade, it will become impossible to grow most heirloom tomato varieties. Few have the level of blight resistance that will be needed to remain a workable crop in the times that are coming. So in the tomato chapter of this book, in addition to all the information needed to choose and grow the most flavorful varieties, I include a comprehensive section on managing late blight in the organic garden as well as a comprehensive list of all known late blight resistant tomato varieties. I also issue a call to arms to all lovers of heirloom tomato flavor to save these wonderful flavors by breeding new late blight resistant heirloom-style varieties of our own—that is, to create the heirloom tomatoes of tomorrow. And I provide all the information about tomato genetics and breeding necessary to facilitate this effort.

The final chapter of the book focuses on seeds, and includes sections on the Do-It-Yourself Seed Bank, on preparing seed for long-term storage, on dehybridizing hybrids, on creating your own modern landraces, on genetically rejuvenating heirloom varieties, and on breeding crops for organic systems.

The Tao of Vegetable Gardening is largely organized by important Taoist principles. The passages in italics at the beginning of most chapters are from *Tao Te Ching*, the twenty-five-hundred-year-old classic of philosophical Taoism that is the most translated book in the world other than the Bible. These versions of the passages are from my own translation *Tao Te Ching: A Window to the Tao through the Words of Lao Tzu*. The Taoist stories in *The Tao of Vegetable Gardening*, also my own renditions of Taoist classics, are from my anthology *Taoist Stories: A Window to the Tao through the Tales of Chuang Tzu and Lieh Tzu*.

The Tao of Vegetable Gardening is a presentation of gardening through Taoism and an illustration of Tao through gardening.

Heaven covers. Earth supports.
Heaven and Earth endure.

In living, honor the land. In thinking, be deep.
In speaking be truthful. In working, be skillful. In dealing
with others, be compassionate. In giving, be generous.
In ruling be fair. In acting, be timely.

道

CHAPTER ONE

Honoring the Land

Gardening in Nature's Image—But Which Nature and Which Image?
Has Nature Thought of Everything? On Being a Member of
a Keystone Species. Organic and Beyond.

Gardening in Nature's Image—
But Which Nature and Which Image?

It's useful for beginning gardeners to look to more experienced gardeners for ideas and methods. We humans have been gardening for only about ten thousand years. We are beginners. Leaf-cutter ants, on the other hand, have been gardening for fifty million years. So what can we learn from these more experienced agriculturalists?

Leaf-cutter ants comprise hundreds of species that live in the New World tropics (including some areas in the southern United States). The leaves of most tropical plants are protected against herbivores by noxious chemicals. The ants harvest pieces of leaves, carry them to their burrows, chew them into a pulp, then use the pulp to grow a fungus. The fungus detoxifies the leaves; the ants eat the fungus. Different species of ants cultivate different species of fungi. The food fungus cannot live without the ants nor the ants without their food fungus. Not only does each ant colony depend on just one species of food fungus, it also cultivates only one line of that species. A young ant queen carries some of her line of fungus with her when she establishes a new colony.

Another fungus, *Escovopsis*, is a weed in the ant gardens. It also eats the food fungus. If ants are removed, the gardens are overrun by *Escovopsis* immediately. Ants control the *Escovopsis* both by physically weeding it out as well as by using chemical fungicides, some produced by their own glands, some produced by a species of bacteria that grows on their bodies. (R. Ford Denison, author of *Darwinian Agriculture*, suggests that the bacteria probably produce the fungicide in order to compete better with a yeast that also grows on the bodies of the ants.) The fungicide affects both the food fungus and the weed fungus; however, it affects the weed fungus more. The ants also apply the fungicide specifically so as to control the pest and minimize damage to the food crop.

Yet another species is part of the ants' farming pattern. A nitrogen-fixing bacterium also grows in the gardens and provides some of the nitrogen for the colony. So leaf-cutter ant agriculture involves one food crop and at least four other species. Agricultural techniques the ants practice include growing food indoors, bringing nutrients for the food crop from the larger area surrounding the nest to concentrate in the smaller garden, intensive growing of food crops, monoculture, processing or chopping up the growing medium to make it a better medium for the food species (a kind of tilling), weeding, fertilizing (with feces), applying fungicides, and using nitrogen-fixing bacteria.

It's a fascinating bit of agriculture and ecology. However, the bottom line is that the ants depend entirely on one variety of one food crop, eat high on the food chain, and farm exclusively using continuous monoculture supported by the massive use of fungicides. These practices are often charged with being unsustainable when humans do them. But it's difficult to argue unsustainability with a pattern that has persisted for fifty million years.

When we are exhorted to garden or farm in nature's image, the practices being encouraged most commonly include using perennials rather than annuals, growing plants in mixed polycultures instead of monocultures, increasing and protecting biodiversity, and refraining from using pesticides, fungicides, and herbicides. Ant agriculture violates all these principles. Human agriculture also often violates one or more of them. But are these principles in fact what nature does?

Nature invented annual plants, plants that complete their life cycles within a year; biennials, those that require two years; and perennials, those that take longer. All of them are widespread in nature.

Monocultures are not rare in nature, at least not the kind of monocultures typical of an ordinary home vegetable garden. When I plant a row or bed of tomatoes there are always some weeds. In addition, there are microorganisms of many kinds as well as earthworms, sow bugs, and other critters. There is a monoculture only if you ignore the weeds and consider just the tomato patch and not the entire rest of the garden, which is full of many other things. In addition, the monoculture is temporary, just for a few months. Agribusiness farms may practice monocultures for year after year on thousands of acres. But this book is about gardens and gardening, and the issue is whether to plant small temporary monocultured patches within our polycultured gardens or try to multicrop on even the smallest scale, that is, plant multiple crops in every bed or patch.

Small temporary monocultures are common in nature. When gophers excavate their tunnels and make mounds of soft, loose soil, what germinates and establishes itself on those mounds often tends to be all one species initially, depending on what seeds happened to be in the soil seed bank and which kind of seeds germinate and grow well at the particular time the gopher created the mound. Similarly, after a fire certain pioneer plants colonize the burned-over land, and there can sometimes initially be nearly pure stands of some of these species. Ants that tend aphids—that carry them around, place them on succulent plants, protect them from predators, and harvest their honeydew excretions—don't also tend caterpillars and beetles. They tend just the monoculture of aphids. If I till a section of my garden and leave it unplanted for a while, the weeds that come up sometimes create near-monocultures and sometimes not, depending on exactly what went to seed in that area recently, what seed is in the soil seed bank long-term, and the time of year. If I till certain areas in early spring and leave them unweeded, I may get a near-monoculture of tansy via the germination of seeds that date back to when the field was an abandoned grass seed farm infested with tansy from a nearby pasture. If instead I till in mid- to late spring I may get a near-monoculture of lambsquarters because I let some go to seed in that area recently, and the timing is perfect for it to germinate and grow rapidly.

Larger and longer-term near-monocultures are also common in nature. Redwood forests have very low species diversity, for example. Look at the classic pictures of redwood forests. The trees are all redwoods, and there is little in the way of understory vegetation. The wild relatives of our major small grain crops tend to grow in nature as near-monocultures.

Using pesticides is also not a uniquely human invention. Many plants produce chemicals designed to repel, discourage, damage, or kill any creatures—from insects to elephants—that dare to try to eat them. Many microbes produce chemicals that kill or deter the growth of other microbes.

Using herbicides is also not uniquely human. There is not much that can grow under walnut trees. The trees produce substances that inhibit the growth of most other plants. I once sowed mustard seed in an area where I had just harvested some barley. The mustard seedlings were all distinctly stunted and wrinkled and sick looking. I transplanted some of the seedlings to a different spot in the garden. They cheered right up, recovered within a few days, and grew happily. Many grasses produce herbicides to deter the growth of competing plants, and these herbicides can persist in the soil for some time after the grass dies or is tilled under.

Nature is extravagant. She has already thought of annuals and perennials, monocultures and polycultures, pesticides, fungicides, and herbicides. It is common to justify using or not using various of these methods by pointing to some specific, cherry-picked example from nature. I suggest we be more honest with ourselves and cease the cherry-picking. Whether we choose to use or forgo any of these methods we need reasons other than pointing to nature. This doesn't mean that we cannot legitimately decide to choose or to avoid certain methods. It simply means that we cannot use nature as our justification. We need to think deeper.

Has Nature Thought of Everything?

Are there any methods or technologies we humans have invented that are not found anywhere else in nature? And if so, does the fact that we are the lone wielders mean that these technologies are somehow evil and should be avoided?

I can think of only a few technologies we humans invented that have not already been evolved or invented and used first by other creatures. We are not the only makers and users of tools or weapons. Chimpanzees throw rocks and branches at snakes as well as modify twigs in ways that make them ideal tools for fishing for termites. Sea otters use rocks to bash open clams. Some creatures sense magnetic fields and electrical currents and use this ability for direction finding or locating food. The electric eel has battery-like structures that allow it to generate such a large electrical jolt that it can stun its prey or even a large person. Sonar is nothing new to the bat. The brain itself is a kind of computer. Many bacteria and some plants are experts at swiping one or a few genes from quite unrelated species, that is, doing genetic engineering. But there are some major things we do that, as far as I know, no other plant or animal does.

First, we control fire deliberately. Other creatures make use of fire. Some plants are adapted to growing in fire-prone areas. Some plants produce seeds or cones that germinate only after they have been scorched by a fire. But other creatures don't set or control fires. Second, we cook our food.

Third, we make and wear clothes. Fourth, we trade. Other creatures may engage in gifting, but nobody else seems to do the kind of deliberate swapping humans do. Fifth, we are also the only inventors and users of the wheel and axle in all of its expressions from cart to car. And sixth, while many creatures can swim, as far as I know we are the only ones who build boats.

Finally, while we weren't the first to invent agriculture itself or domestication, fertilizing, pesticides, fungicides, herbicides, or genetic engineering, we do appear to have first invented six important agricultural methods. Beavers share with us the technology of making dams and raising water levels, but we are, to my knowledge, the only creature that deliberately creates drainage and lowers water levels. We are also the only farmers who irrigate, or who practice crop rotation. In addition, while some insects such as ants and termites farm fungi, and ants herd aphids, we are the only creature that farms green plants, or that does grafting, or that farms outdoors.

What should our attitude be toward these uniquely human methods? Should we stop controlling fire, cooking, wearing clothes, trading, or farming with green plants outdoors just because we know of no other creature who evolved or thought of these methods?

More to the point, why should the inventions of every other creature on the planet be considered natural and honorable, and only those of humans be deemed unacceptable? Is not a human as legitimate a creature as an ant?

We evolved on this planet just as the ant did. We belong here. Earth is our place, our home. We are a part of nature. If the ant is natural, so am I. If the ant's garden is natural, so is mine.

On Being a Member of a Keystone Species

Every living creature on the planet influences the environment around it and the welfare of other creatures. Even a bacterium or fungus growing in a clod of soil removes nutrients and makes less of them available for other creatures as well as produces other products that help or harm those around it in myriad ways. And this is before you even consider that the microbe might have evolved more sophisticated ways of interacting with others, such as being able to nodulate and fix nitrogen for legumes or form mycorrhizal associations with plants, or manufacture antibiotics designed to kill or stunt competitors. Some species, however, make massive changes in their environments, alter the landscape fundamentally, and change the relative abundances of many other species. We call these creatures *keystone species*. A pair of beavers moves into a stream flowing through a meadow, dams the stream, and creates a pond, a higher water table, and marshy or seasonally flooded areas that are ideal habitat for the willows and poplars that are favorite beaver foods. It is in the very nature of the beaver to build engineering projects that cause such large transformations. The pond and marsh the beavers create are not better or worse than the creek and meadow they destroyed. The beavers' engineering creates a better environment for beavers as well as for all other species that thrive in ponds or marshes. Inevitably, this means a worse environment for other species, those that prefer better-drained soil and uplands. The beaver is a keystone species. So are we.

Look at our hands. They are designed to make tools, build things, change the environment. It is part of our essential nature. We are inherently members of a keystone species. We need not apologize for it. However, not all changes we can make are advantageous, even to ourselves. And the more powerful any technology, the more care we need to exercise in deciding whether or how to use it. Humans have no choice as to whether we will be a keystone species or whether we will have a large influence on our environment, especially now that there are seven billion of us. We do, however, have lots of choice as to what amounts and kinds of influence we will have, and on which parts of our environment.

I think the conversations and arguments about gardening or farming in nature's image, about environmentalism, about organic versus conventional agriculture, about reducing our energy consumption or carbon dioxide production, about trying to have a minimal environmental footprint—reflect the fact that we have realized that we are a keystone species, one so powerful that we have the potential to destroy ourselves and our planet utterly. We have moved beyond the first blush of innocence, when we used every technology we invented and only later discovered the downsides. We have learned that we can't do just one thing, that everything is connected, that every technology we use has myriad side effects, some inevitably unpredicted and some inevitably undesirable. These models are our attempts to create frameworks for guiding us in choosing, using, or restricting our technologies. They represent our first efforts at becoming more responsible members of a more responsible keystone species.

Gardening or farming in nature's image is one model designed to guide our choices. As I have explained, however, gardening or farming in nature's image offers useful guidance only if we cherry-pick our examples to fit preconceived ideas.

In addition, the purpose of a garden or a farm is to produce more and better food for humans than nature would if left to her own devices. Each species in a patch of nature has evolved primarily to survive itself. Only our domesticated plants and animals have been selected or evolved primarily to produce large amounts of prime food for humans. Our domesticated plants spend such a large part of their energy on producing food for us that few of them can even survive in nature. They can't compete with wild plants. If we really could farm in nature's image, we would need to eat wild plants and animals, not any of our much more human-food-productive domesticates. That would be possible for a much smaller population of humans living in certain areas. It could not provide for the food needs of seven billion people.

Another possible model is one of simply trying to have as little impact as possible. "Take Only Pictures. Leave Only Footprints." That is what we are encouraged to do when visiting wilderness areas. However, the minimal impact model works best in places we are just visiting, in environments we can afford to refrain from using to support ourselves in any way. Gardens are where we live and support ourselves.

Gardens can have appearances ranging from very tame and tended to quite wild and unruly. But whatever the superficial appearance, productive gardens are always highly modified in the direction of making them do a better job of producing food for us than the land was doing before we built the garden. Any attempt to have as little impact as possible turns the garden back into land that produces less human food.

Organic gardening is another model for making decisions about what technologies to use. Organic gardening largely works and is easy to do on the scale of the home garden. It isn't a perfect pattern by any means, but it covers many of the issues we need to consider and makes workable decisions about most of them. In the next and final section of this chapter I consider the basics of organic gardening as well as some of its limitations.

Organic and Beyond

Essential to the practice of organic gardeners and farmers is honoring the land. It's an emotional and spiritual attitude—a mix of love, humility, gratitude, awe, and reverence. *Land* refers to not just the bit of soil to be turned directly into garden, but the surrounding land in the neighborhood and beyond, the ecosystem, all its plants, animals, and other creatures, and the rest of the planet.

We try to leave room on our land for wildlife, wildflowers, and natural biodiversity of all kinds. If we are living on a small lot in a city or suburb, we might turn every bit of land we have into food crops in lieu of the standard and expected lawn. One less lawn will never be missed, we figure. However, whenever we have more land than we need we try to create or preserve niches for native plants and animals and for the full range of biodiversity that is adapted to our land. We especially value and try to preserve any rare ecosystems there might be on our land. We do not plow, till, plant, or mow every bit of land just because we can. We exercise restraint. And we are especially careful not to plow steep land or other areas that might be subject to erosion.

Organic gardeners do not use synthetic fertilizers. These are highly soluble and tend to run off and cause pollution. In addition, soluble fertilizers impede or sabotage the nitrogen-fixing bacteria and mycorrhizal fungi that are essential parts of providing plants with nutrients in natural soils. Instead, if our soil is low in elements such as calcium, phosphorus, potassium, or sulfur we amend it by adding the appropriate amounts of the appropriate kinds of ground rock, such as limestone, dolomite, gypsum, or rock phosphate. These powdered rocks are non-caustic and naturally slow-release. We provide our garden plants with nitrogen by incorporating into our gardens nitrogen-containing organic materials such as tilled-under cover crops or green manure crops, compost, manure, pelleted chicken manure, fish meal, seed meals, or grass clippings. We depend on naturally occurring soil microorganisms to break down the organic materials and release the needed nutrients to the plants slowly and in a way that usually does not cause an excess of soluble nutrients that can interfere with the activities of nitrogen-fixing bacteria or mycorrhizal fungi or that can leach off as pollution.

We pay a lot of attention to the organic matter content in our soil, as it not only provides much of our soil's fertility but also contributes to good tilth—that soft, fluffy texture that makes a soil easy for seeds to germinate in and emerge through and plant roots to grow in.

Organic farmers usually practice crop rotation, and organic gardeners usually also do so to the extent that they are able, given the more limited space in which they work. Crop rotation is one aspect of establishing and maintaining soil fertility, but it is also important for preventing and limiting soilborne diseases (such as *Fusarium* in tomatoes). In the small home vegetable garden it isn't usually practical to rotate all crops. But it is highly advantageous and often essential to rotate at least the members of the Nightshade family (Solanaceae): tomatoes, potatoes, peppers, eggplants, and tomatillos. I plant my potatoes in one unirrigated patch and my tomatoes, peppers, and tomatillos all together in the irrigated garden, and keep track of where each is from year to year. I avoid planting nightshades in that spot again for the next two years. In larger gardens such as my seed production fields we rotate everything so that no crop grows in the same place two years in a row. However, in my smaller home food garden I'm careful to rotate the nightshades, but I don't worry about the rest.

Organic gardeners and farmers do not use synthetic herbicides. There are supposedly organic herbicides such as corn gluten meal, which is used as a preemergence herbicide. (This *gluten* refers to a protein fraction of corn; it is not the same as the specific protein in wheat that causes the problems for those with celiac disease.) Corn gluten meal is expensive and not very effective compared with commercial herbicides. Vinegar or concoctions made with vinegar can burn off the tops of the weeds, but do little damage to the roots. Basically, there are no effective organic-approved herbicides.

Back in the days when I was living in town in Corvallis, Oregon, I had trouble with asthma. Whenever anyone sprayed commercial herbicides, even just from a small hand sprayer a full block away, if the winds were such that I caught the drift, I had an asthmatic reaction. Whether I was reacting to the active ingredient in the herbicide or to the oils and soaps included to make the herbicide stick to the plants isn't obvious.

Frogs are known to be killed by herbicides. University of Pittsburgh scientist Rick Relyea found that, when applied in recommended amounts, Roundup herbicide killed 79 percent of terrestrial frogs in the treated area within one day. And in amounts that would typically reach waterways when applied on nearby land Roundup killed 98 percent of all tadpoles within three weeks. It was one of the so-called inactive ingredients in the herbicide, a surfactant designed to help the herbicide penetrate plant leaves, not the active ingredient, glyphosate, that killed the tadpoles. Frogs are undoubtedly not the only animal harmed by herbicides. They are just the obvious ones to investigate since their skins are thin and permeable. In addition, the population of amphibians has been declining dramatically in recent decades, and our agricultural chemicals are high on the list of possible causes.

I suggest that when the "inactive" ingredients in herbicides cause asthma attacks in people and the death of amphibians the word *inactive* isn't quite appropriate. Yet herbicide manufacturers are not required even to list the inactive ingredients on the labels. Oil and soap alone can be expected to have widespread biological effects on many creatures that are exposed to them. In addition, since herbicides are usually used as sprays, they usually spread beyond the area where they are applied and get into ground and surface waters. Plants in my first garden in Oregon were killed when the county sprayed the edge of the road more than fifty yards away.

Like most organic gardeners, I control weeds by tilling, hand hoeing, pulling, or mowing, not with herbicides.

Organic gardeners don't use synthetic fungicides. There are other fungicides, such as copper-based fungicides, that are traditionally used by organic gardeners and that are officially sanctioned by the USDA National Organic Standards Board, the organization that determines acceptable practices for those who sell food that is labeled organic. I don't use any fungicides, whether they are considered acceptable for organic growers or not. I don't want to kill off the fungi in my garden soil that form mycorrhizal associations with plant roots and help them take up phosphorus and other nutrients. Fungi are also critically involved in breaking down organic matter in the soil so as to release the nitrogen and other nutrients plants need. A healthy, vibrant organic soil should be full of fungi. I prefer to deal with fungal diseases by using or breeding resistant plant varieties and using crop rotation and disease-restricting watering patterns and other practices. (See the section in the tomato chapter on dealing with and managing tomatoes so as to prevent or minimize problems with late blight.)

The commercial fungicides can also kill frogs. Taegan McMahon, a University of South Florida scientist, found that the most widely used fungicide in the country, chlorothalonil (Bravo, Echo, Daconil), killed 90 percent of every frog species they tested when subjected to the doses you would expect in waterways near farm fields, lawns, or golf courses where the fungicide is applied. I don't know whether the organic-allowed fungicides are any better; I have not seen them tested. I suspect anything that is actually effective as a fungicide will turn out to harm frogs and other creatures once someone gets around to studying the question. That is, I think whatever impressions we might have as to the organically approved fungicides being more environmentally benign than the commercial types may mostly reflect a lack of information.

Sometimes seed companies treat seed with fungicides. This treated seed is not allowed in certified organic plantings, and I think it is strongly to be avoided. Treated seed allows seed companies to sell us seed of varieties that are not resistant to common soilborne fungi such as *Fusarium* or *Pythium*. This means if we save seed, the seeds we save won't germinate well in our ordinary soil without the special seed treatments. The fungicides used in the seed treatments are dangerous, and applying them really isn't a home process. So if you buy seed of varieties that can't germinate without the seed treatment, you have a variety you cannot propagate yourself. Some seed companies have tried to force gardeners to accept such dependency-creating varieties and seed treatments by offering all their seed in only treated versions. *Organic Gardening* magazine, when it was at the height of its power and influence, put an end to that practice by refusing to list, acknowledge, or include in its articles any varieties that were available only as treated seed. In addition, USDA rules for organic growers (which must be followed if you sell crops as organic crops) do not allow commercial fungicide seed treatments. Most seed companies that produce treated seed these days also offer untreated seed. (Where they don't, I suspect the variety is too susceptible to soilborne fungi to germinate without the treatments, so it should be avoided anyway.)

I suggest that even if organically acceptable seed treatments are developed we still shouldn't use them. If we need to use any kind of seed treatment to get a variety to germinate, this makes it harder for us to save our own seed. Being able to save our own seed lowers our gardening costs, allows us to breed and adapt varieties to fit our particular ideas, needs, and growing conditions, and enhances our independence and resilience.

Organic gardeners don't use synthetic pesticides. However, there are a number of natural pesticides that are sanctioned for use by organic growers. Just because a product is natural doesn't mean that it is safe to use or has no potential for environmental damage, however. The ricin from castor beans is a natural product, but you can kill yourself very quickly by eating it. Are the organic-approved pesticides safer to use than the synthetic ones? Do the organic-approved pesticides do less damage to non-targeted creatures and the rest of the environment? There is an excellent product-by-product study and evaluation of the efficacy, safety, and potential for environmental damage of each of the common synthetic as well as organically sanctioned pesticides (and herbicides and fungicides) in *The Truth About Organic Gardening: Benefits, Drawbacks, and the Bottom Line* by Jeff Gillman. I highly recommend this book to anyone who is considering using herbicides, fungicides, or pesticides of any sort, organically approved or not. The simple answer is that some of the organically sanctioned pesticides are actually more dangerous and have more deleterious environmental effects than some of the synthetic ones. It all comes down to exactly which product you are discussing.

My own approach is simple. I don't use any pesticides, whether they are generally considered acceptable in organic gardening or not. Many pesticides kill the natural predators of pest insects more effectively than they kill the target insect. Once the predators are gone, the population of the problem insect rebounds; they then become more numerous than ever since their natural predators are gone. Even if the pesticide doesn't kill the natural predators directly, it disrupts their food supply and may lower their population levels. Predatory insects are higher on the food chain and are less numerous than the insects they eat, and their populations are generally more subject to disruption and are slower to recover. I leave my insects alone so as to keep my insect predator populations undisturbed and as effective as possible. I don't even pick insects off plants when I see them, preferring to operate at a system level instead of doing that sort of micromanagement. I instead focus on using or breeding crop varieties that are so vigorous and fast growing that they aren't affected much by insect attacks.

Organic gardeners and farmers do not use GMO (genetically modified organism) varieties.

Neither the traditional organic approach nor the legal USDA organic standards for those who sell organic crops say anything about using hybrid seed. This is a failing, in my view. Most hybrids are created so as to allow the seed company to control the seed and prevent seed saving, not because the hybrids are better than the best equivalent open-pollinated (non-hybrid) varieties. (There are exceptions, however.) Hybrids, plant patenting, and plant variety protection (PVP) are all physical or legal mechanisms aimed at preventing us from fully owning the seed we paid for. In effect, we are only leasing certain temporary rights to the seed. And the seed, because of the extra work of making the hybrids, as well as the monopoly involved, is much more expensive than conventional seed. I think all hybrids, patented varieties, and PVP varieties should have been excluded from the USDA's organic standards and rejected by organic growers. And in fact many organic gardeners and farmers do avoid all hybrid, patented, or PVP-protected varieties.

I think being able to save our own seed is an essential part of a healthy, resilient gardening and farming pattern. If we can't save our own seed, we also can't do our own plant breeding—that is, we cannot develop the varieties that are ideally adapted to our specific needs and local growing conditions. I plant no seed I don't fully own except in trials, or in order to dehybridize it or to use it in the development of "real" varieties, that is, open-pollinated, public domain varieties.

CHAPTER TWO

Honoring the Essential Nature of the Plants

*Sun, Earth, Air, Water, Warmth. What Can We Grow? Expected First
and Last Frost Dates. Sun and Shade Tolerance. Some Like It Hot;
Some Like It Cold. When to Plant Everything. Planting Guide.*

Once upon a time in ancient China the Yellow Emperor got word that
the great sage Tai Kuei had taken up residence on nearby Chu Tzu
Mountain. "A true sage is just what I need right about now," the Yellow
Emperor said to his ministers. "It seems to me that the longer I rule, the
less I know about it. I'll find this sage and ask him how to govern the
empire." So the Yellow Emperor set forth in his carriage for Chu Tzu
Mountain along with his ministers and advisers and military escort.

When the Emperor and his retinue came to the foothills of the
mountains they found themselves in deep woods crossed by many trails,
and they lost their way. Presently they came to an open meadow where a
young boy tended a herd of horses grazing near a creek that ran through
the meadow. The boy appeared to be about eight years old. He lounged at
his ease on the back of a horse he rode with no saddle or lead.

"Hey, you, Boy!" the Yellow Emperor called, leaning out of the window
of his carriage.

The boy rode over.

"Do you happen to know the way to Chu Tzu Mountain?" the Yellow
Emperor asked.

"Certainly," the boy responded, and he gave simple, exact directions.

"Thank you," said the Yellow Emperor. "Do you happen to know the
way to the residence of Tai Kuei?"

"Certainly," said the boy, and gave the directions. Then he added, "But
he isn't at home right now."

"Well," said the Yellow Emperor. "You are certainly an astonishing boy!
Tell me . . . do you happen to know how to govern the empire?"

"Certainly," said the boy. "You govern the empire exactly the same way
you care for horses."

"How is that?" the Yellow Emperor asked.

The boy responded. "Caring for horses is easy. You simply get rid of anything that is harmful to the essential nature of horses."

The Yellow Emperor sat there for a moment, stunned. Then he got out of his carriage and prostrated himself before the boy on the horse.

The boy smiled and nodded slightly in acknowledgment. Then boy and horse ambled off to rejoin the herd.

Sun, Earth, Air, Water, Warmth

All green plants use the sun to provide the energy they need to transform substances they take from earth, air, and water into more plants. When we plant our seeds or transplants in soil outdoors, that takes care of the earth and air. Moisture in the ground, rain, or our irrigation provides the water. Plants also need the right amount of warmth. They can do photosynthesis and perform their complex conversions only within a certain range of temperatures, which varies with the species and variety. So we plant varieties that are suitable for our region. And we plant them at the right time of year so as to provide the right amount of warmth as well as to give them enough time to produce their crops. That's all there is to growing vegetables. The rest is details.

If we just go out in the backyard, make a slit in the grass sod, and sprinkle some vegetable seeds on top of the soil, they are unlikely to be able to germinate, establish themselves, and grow well enough to do anything useful. The grass is already established, and its roots outcompete the new seeds for nutrients and water from the soil. In addition, the grass will regrow quickly and deprive the seedlings of sun. Usually, to get new seed to germinate and grow we need to first remove existing competing vegetation by digging, tilling, or plowing it under. With a small garden bed we might actually remove the sod completely with a shovel and a heavy hoe and put the sod in the compost heap. Then we dig the bed to loosen the soil, often adding various amendments at this time. This soft soil makes it easier for our plants to establish themselves. With bigger gardens and fields we usually plow or till twice. First we till to break up the existing vegetation and turn most of it under. Then we wait two or three weeks for the material we have turned under to die. Then we plow or till a second time to turn under the bits of preexisting vegetation that are near the surface and that have already begun to regrow. We usually plant our own crop right after that second tilling or plowing.

Since our domesticated plants invest so much of their energy in producing succulent food for us they are usually not competitive with wild plants or weeds that use all their energy for surviving and reproducing themselves. Weeds love vegetable gardens. If weeds are allowed to flourish they will be harmful to the essential nature of our vegetables on multiple counts. They will rob our vegetables of soil nutrients, water, and sun. So we manage tilling, planting, and timing so as to favor our crops over the weeds. In addition, we physically weed out weeds by tiller, hoe, or hand.

Our domesticates are from different places on earth that have different kinds of soil. Our own native soil may need some amendments and fertilizer to make it optimal for our vegetables. We try to make conditions as favorable as possible so that the vegetables will grow at optimal rates. Poorly grown vegetables tend to be undersized, tough, stringy, or bitter.

Our garden plants will also compete with one another if they are too crowded, and the result is stunted, unproductive plants that produce inferior food or don't produce at all. So we thin to the optimal density of plants.

If we plant our seed too shallowly, the seeds might not get enough moisture to germinate, or they may begin to germinate, then dry out and die. If we plant too deep, our seeds either will not germinate or will be unable to emerge through such a deep soil layer to reach the light after they germinate, and so will die. Planting big and little seeds presents different kinds of problems. I cover how to plant each in the chapters on peas and beans and on greens, respectively.

We usually sow more seeds than we need for the amount of space, often four times as much seed or even more. This is because we expect critters to get many of them. Traditionally it was said: "One for the mouse, one for the crow, one to rot, and one to grow." Also, some seed might be weak and not emerge. And some might be too shallowly or too deeply planted. After the seedlings emerge and establish themselves, we thin the plants to the optimal density, usually with a hoe. Most seed company listings will tell you the opti-

mal final spacing for the plants. It depends on the exact variety being grown and the fertility of the soil, but the listings are usually close enough as a first approximation. If the variety makes bigger plants than most varieties of that crop, give it more space. If our soil is a bit less fertile than typical garden soils, we give the plants a bit more space. If the soil is tilled relatively shallowly, we give the plants more space than if they are in a double-dug bed. If we prefer to water less frequently we similarly give the plants more space.

It's best to plant enough seed so we end up with more plants than we need. This allows us to eliminate the weaker plants and keep only the best. This gives us more and better produce. But in addition, it allows us to maintain or improve the variety if we save our own seed. If we keep seed from a mix of plants that includes the inferior plants, our variety will deteriorate year by year. Good seed saving requires maintaining or improving the quality of the variety every year, and that requires planting more seed than we need so we can select and keep seed from just average to superior plants, not inferior ones. We do not want to sow too many seeds, however. That not only wastes seeds, but also makes thinning more laborious.

What Can We Grow?

Some crops and some varieties need more warmth or a longer frost-free growing period than we have. In New England or Canada or Oregon we are just not going to be able to grow oranges outdoors. And nowhere in temperate North America can we grow coffee or chocolate. You can figure out what crops you can grow by seeing what your neighbors grow, asking experienced gardeners in your area, or consulting regional seed company catalogs. I list regional seed companies in the appendix. (A regional seed company is one that focuses on their own region, even if they sell nationally. Everything they carry will grow in their region.) You can also consult regional gardening books, or check out the website of your local land grant college's agricultural extension service.

Farmers markets—the kinds that carry only local produce—can also be a good source of information for gardeners. You don't have to do too much wandering around farmers markets before you have a pretty good idea of what you can grow. Whenever you see something you think you might like to grow, ask the people tending the booth whether they grow it outdoors or in a greenhouse or hoophouse.

Another approach is to use the USDA Plant Hardiness Zone Map that you can find at www.planthardiness.ars.usda.gov. The USDA map gives you a number that is determined by the average coldest winter temperature in your area. So if you are zone 8b, for example, as Corvallis is, you presumably can grow any crop designated as for zone 8b and lower, but not 9a or higher. The total amount of summer heat and your dates of last frost tend to track with the hardiness zones, though better in some parts of the country than others. The correlation tends to break down in the western half the country because of the moderating effect of the Pacific in summer. It also breaks down wherever there are mountains, which is most of the West. The zone is lots more relevant when we are planting trees and bushes and are worried about whether they will freeze outdoors than when we are considering whether we have enough summer heat and growing season to grow a 120-day winter squash. And without more

specific information about when our last expected frost date is, we don't know when to plant the squash, even if we have a long enough season for it. We need to know the exact average date of the last expected frost in our region.

Expected First and Last Frost Dates

Some plants are tender, that is, they cannot tolerate a freeze, and must be planted after all danger of freezing weather is over. Others can take a little freezing, or a lot of freezing, so they can be planted anywhere from a bit before to even weeks before the average date of last frost. Basically, we need to know when to expect the last frost in our area in order to know when to plant everything.

In the past we usually found out our expected average last frost date in spring from local gardeners, local seed companies or nurseries, or local gardening books. These days you can look up such things on the Internet. For example, I can swipe across the home key area on my Google/Nexus 7 tablet to trigger the voice search and say, "Okay, Google. What is the average last frost date for Corvallis, Oregon?" (I could type it in, but, boy, is it a lark being able to just ask.) The first entry is "Frost Dates for Oregon from Victory Seed Company." When I click on the link I get a page listing the average last spring and first fall frosts for about forty cities in Oregon, with Corvallis listed in bold. May 11 is the date of the average expected last frost in spring, it says. And October 8 is the date of the average first frost in fall. The Victory Seed information covers the entire United States but not Canada. A similar Google search for cities in Canada taps into Canadian seed companies and data.

The date of average last frost is the average date for the reference city or weather station. You should add or subtract a little to correct for your exact location as well as for your microclimate. My home garden is in a lower area that is a frost pocket, for example, so I add a week to get from May 11 to about May 18. Then, in addition, I add a bit more to get the expected date beyond which I am unlikely to get any frost almost any year. I want to plant my tender vegetables after the last real frost *this* year, not after the average frost date. This is because half the years the average date is going to still be too early for planting tender vegetables. I take the end of May as the end of the period that I am likely to get frosts. That is an approximation that you then modify from your own experience. In my experience, I'm pretty safe planting tender plants in my garden on or after June 1. I am very unlikely to lose them to frost. If I plant in the third week of May I would get a head start many years but would lose the plants about one year out of every three to five years, I'd say. So whenever gardening instructions say "Plant after all danger of frost is over," for my particular garden in its frost pocket microclimate I use a date of June 1.

October always finds me keeping careful track of weather reports, watching out for the first frost. Some crops can be overwintered in the garden, but many crops must be gotten in before first frost. In particular, I need to finish harvesting all the winter squash before first frost, as being frosted even lightly damages the storage longevity of the squash. A serious freeze damages them so badly that they just rot instead of store.

The difference between the average last frost date in spring and the average first frost date in fall is your average frost-free period. We often use that number to figure out whether our growing season is long enough to grow a particular vegetable variety.

Sun and Shade Tolerance

Some vegetables need more sun than others. In general, where the crop is a fruit, it needs full or mostly full sun. Where the crop is leaves or stems it is usually able to produce well in partial shade. Root crops tend to be somewhere in between greens and fruits in their need for sun. But the species matters too. Even the variety matters. Some varieties of a crop may be able to perform better in shade than others. Seed catalogs often give us information about the sun needs of the crop in general, but usually don't identify varieties of a crop that have more or less shade tolerance than most varieties of the crop. You have to just try and see.

Not all shade is created equal. Dense shade cast by a building or by certain trees blocks the sun much more completely than the light shade cast by wispy foliage of other kinds of trees.

Not all hours of sunlight are created equal. If your garden is shaded at the beginning or end of the day, this may not affect your garden much. That's because at the beginning and end of the day the sun's rays are less direct and less effective with respect to allowing the plants to do photosynthesis than when the sun is more directly over the plants.

You can grow most greens fine on just a few hours of sun a day. But growing crops that make fruits such as squash or tomatoes is more difficult. However, for the shady garden transplanting can be a game changer. One of my friends has a garden that is buried in a deep woods and has only a few hours of sun per day, just the few hours when the sun is nearly directly overhead. Yet he manages to grow corn and squash, even the big full-season winter squash 'Sweet Meat—Oregon Homestead'. The trick is he transplants everything that would ordinarily need more sun than he has, even the crops that are normally direct-seeded such as the corn, peas, and beans. By starting the plants in a greenhouse (with plenty of sun) he gives them enough of a head start so that they have time to produce before the end of the season.

In temperate climes, as the summer progresses the rays of the sun become more direct. This means that if your garden is too shady to produce certain crops in early summer it may be able to produce them later in the season. If your garden is too shady to plant bush beans or corn when planted at the usual times, try using a shorter-season variety and just planting it later.

Some plants don't like the heat and grow better in cooler spots, which can be the spots that are partially shaded. I've found that greens and peas planted in midsummer may do better in the partial shade between rows of corn than in full sun. When planted in early spring, they definitely prefer full sun in my climate.

An hour of sun in Texas and an hour of sun in Canada are also not created equal. If you live in the southern parts of United States, that swath across the country from Southern California through Texas to Georgia and Florida, the sun your garden gets is more direct than what we experience in the more northern latitudes. If I can grow a given variety only with full sun in Oregon, you may be able to grow it fine in partial shade if you live farther south. In fact, many varieties that prefer full sun in the rest of the country may do better for you with partial shade.

Some Like It Hot; Some Like It Cold

Some plants don't grow very well in cool weather, even if they aren't harmed by actual frost. We usually plant peas earlier than corn, for example. Both can take serious freezes in the seedling stage. But peas actually prefer relatively cool weather, and grow pretty well even when it's cold. Corn really wants it to be warmer. If we plant the corn too early, the seed can sit there in the soil, not growing, until molds or insects get it.

Some of our crops, especially many of the greens and root crops, prefer cool weather so explicitly that we plant them and grow them only in spring and fall, not in the middle of the summer. In addition, some greens bolt if planted at the wrong times. That is, they make flowers and go to seed while still little plants, before they have made a good yield of food for us.

Individual varieties of a crop can have their own patterns with respect to how hardy they are, how cold- or heat-tolerant, and when they bolt. Some varieties of a crop might be intended for fall planting and others for spring or summer, for example. When a listing shows radishes or cabbages being planted in many months, sometimes these are the same varieties, and sometimes they are special spring-only or fall-only varieties. You have to look carefully at the variety descriptions in the seed catalogs.

Growing instructions on seed packets are often misleading, by the way. The seed packet information is usually generic for the crop in general, and describes how the most common varieties of the crop are grown, which may or may not be accurate for the seed of the variety that is in the packet. So, for example, many radishes are planted spring through fall, but not so-called winter radishes. They are never planted in spring. They are planted in summer for fall production. But the growing information on the packet of winter radishes is usually for the more common spring-planted kinds, not the winter radish variety whose seed is in the packet. In most cases you need to consult the seed company's main catalog for growing information on the specific varieties, not just rely on the information on the seed packet.

When to Plant Everything

The accompanying Planting Guide is a general planting schedule I've put together from various sources as well as my own experience. Since it is stated in terms relative to your expected last frost date in spring, it should work pretty well just about anywhere.

If you live in the South, you may plant more crops in fall for winter growing than I do. You might also plant more of the leafy crops in spring and fall or even just in fall instead of growing them straight through the summer.

In places with harsher winters than mine my two pre-last-frost categories of planting in spring may be truncated into one for you; that is, everything planted before your expected last frost date becomes "plant as early as you can work the ground in spring." In addition, you will not be planting so many crops for overwintering.

The names of the crops are in bold when that crop first appears in the yearly schedule. So, for example, we first plant lettuce in early spring, so it's listed in bold there; but we continue planting successions of lettuce, so it's listed in standard type in the other planting windows. Where a crop is planted in spring and fall but not in summer, the name is in italics when the crop is first grown again in the fall.

Tomatoes, peppers, eggplants, and artichokes are tropical plants we always start growing indoors under lights and then set out as transplants after all danger of frost is past. Summer squash, cucumbers, and melons are often transplanted to give us earlier harvests. These are tender plants, and are also set out in the garden after danger of first frost.

Planting Guide

Early Spring (can take freezes)
(March in Willamette Valley)
arugula, cabbage, carrots, chicory, endive, fava beans, garbanzo beans, greens, kohlrabi, lettuce, mustard, parsnip, peas, peas–osu*, radish, spinach

Spring (can take light frosts)
(April in Willamette Valley)
beets, broccoli, cabbage, carrots, **chard, collards, corn, cress,** garbanzo beans, greens, **kale,** kohlrabi, **lentils,** lettuce, **onions, parsley,** peas–osu*, **potatoes, quinoa,** radish, **rutabaga, shungiku,** spinach, **turnip**

After All Danger of Frost Is Past
amaranth (grain/greens), annual herbs, beans, beets, broccoli, brussels sprouts, cabbage, **cantaloupe,** carrots, collards, corn, **corn salad, cowpeas,** cress, **cucumbers, dill, ground cherry,** kale, lettuce, onions, parsley, peas–osu*, **pumpkin,** quinoa, **runner beans,** rutabaga, **soybeans,** spinach, **squash–summer, squash–winter,** tender transplants, turnip, **watermelon, yard-long beans**

Midsummer (some successions; some for fall, winter, or overwintering)
amaranth (greens), *arugula,* beans (bush), beets, broccoli, brussels sprouts, cabbage, **cardoon,** carrots, *chard, chicory,* **chives,** *cilantro,* collards, cress, dill, kale, kohlrabi, leeks, *mustard,* onion–scallions, **pac choi,** parsley, *parsnip,* peas–osu*, radish, turnip

Late Summer (for fall, winter, or overwintering)
arugula, beets, broccoli, cabbage, carrots, chard, chicory, cilantro, collards, **corn salad,** cress, *endive,* kale, kohlrabi, *lettuce,* mesclun, **onion–Walla Walla** (direct seeding for overwintering), onion–scallions, pac choi, parsley, peas–osu*, *radicchio,* radish, *salad greens, shungiku, spinach,* turnip

Early to Midfall (cover crops, overwintering crops)
cover crops, Egyptian onions, *fava beans,* **garlic,** greens, lettuce, *peas,* peas–osu*, spinach

* Pea varieties bred by Oregon State University that carry resistance to pea enation mosaic virus, pea wilt, and powdery mildew; these can be planted spring through summer in the maritime Northwest.

道

Other people are noisy and exuberant. I am silent—like something
that has not yet given any sign—like a baby who has not yet smiled.
Other people have many possessions. I have nothing. Other people know where
they're going and what they're doing. I am ignorant. Other people judge things
and make many fine distinctions. I find everything subtle and complex.
Other people act with purpose. I drift and float. Formless am I—like the ocean—
shapeless, unmoving, unresting. I suck from the breast of the Mother.

To know others is knowledge. To know self is wisdom. To conquer others
shows strength. To conquer self is to be strong. To control others shows power.
To control self is to be truly powerful. To know when you have enough is to be rich.
To know when to act is to act appropriately. To know when to hold your place
is to endure. To live at one with the Tao is to live forever.

CHAPTER THREE

Honoring Your Own Essential Nature

Discovering Your Inner Gardener. Planning
Versus Spontaneity. Structure, Labor, and Freedom.

Discovering Your Inner Gardener

"I work in a basement office with no windows. And the lights are fluorescent lights I can't change or control. Is there any kind of plant that can do well there?" Sometimes a question from a beginner brings back vivid memories—in this case, memories of myself as a young molecular geneticist with my newly printed PhD tucked behind one of my still-wet ears, standing in the middle of a big, beautiful, well-equipped lab in a brand-new building. I had an actual job as a scientist, something almost unheard of for a female in those days. I was head of my own laboratory and research program. I had equipment and funds for more equipment. I had hired my first two technicians, got my first couple of graduate students, and won my first couple of grants. Like most molecular geneticists of that era, I worked sixteen-hour days and I loved it. But as I stood in the lab that day I realized that something was wrong.

I had spent much of my childhood outdoors wandering through fields, woods, and swamps. The time demands of my career as an ambitious molecular geneticist meant that I spent almost every hour of my life in my lab or office. My office had a narrow strip of glass about a foot wide as the sole indication that there was any outside world. The window did not open. The labs were all arranged in the center of the building. They

were large, spacious, and well equipped. But they had no windows.

Feeling drives I did not understand, I bought three potted citrus trees and a book on growing potted citrus trees indoors. That book was the first garden writing I ever read. I put a potted citrus tree at the ends of each of three lab benches in my lab. I didn't know whether the trees would grow under the kind of fluorescent lights we had in the lab. But they did. They grew and thrived. All I did was water them. The Meyer lemon did especially well and flowered long and frequently, filling the lab with its wondrous aroma . . . and its intimation of other possible ways of working, living, and being. It was a start.

Much of my own satisfaction from gardening is interwoven with my delight in producing food. Certainly part of that delight is in producing especially delicious food of kinds and quality way beyond anything I can buy. But I think I would enjoy growing food even if it were exactly the same kinds and quality as what I can buy.

In psychology experiments, birds or rodents who have learned to work for food pellets will continue working for pellets even if their source of free pellets is restored. They eat some of the free food. But given the option of working for the same food, they actually cut back on the free food and spend serious amounts of time working for identical food. They prefer to work for at least part of their food. I understand. Hunger adds to the eagerness of a cat hunting mice or birds. But most cats will spend at least some time hunting if they have the opportunity, even when well fed. Evolution has programmed us creatures to have drives to engage in efforts to feed ourselves and to enjoy those drives and efforts. When the food is free and we are deprived of the opportunity to seek and scrounge and scurry and harvest it ourselves, something is missing.

I think accumulating money, working at ordinary jobs, collecting stamps, cars, marbles, or anything else, and shopping are all often sublimations of our basic hunting and gathering drive. But many of us don't find these substitutes fully satisfying. We enjoy more direct hunting and gathering. Hunting, fishing, and gardening—these are so satisfying that we will do them for "free"—even when we don't really need the food. In fact, we will often pay a good bit to do them. For some of us, the vegetables and fruits we grow represent a better-quality diet than what we can afford to buy. Growing some of our own food is partly a matter of economics. For others, what matters most is the fact that the food we produce is so extraordinary compared with anything we can buy. Growing our own food is more a matter of quality than cost. But often, a big part of what we are after is just the joy that comes from that simple, purposeful, productive labor. We want to work for our food. Part of our essential nature is that we are hunters and gatherers. We want to hunt and gather. We create gardens so as to have a place that is an ideally rewarding place to hunt and gather.

When we first start gardening we often have only vague ideas of what we need emotionally from our gardens. My potted citrus trees in the office blossomed, but never produced any actual food. They just satisfied my desire for green growing things in my life. Later, I found I got the most satisfaction from growing serious amounts of food as well as food of the highest quality. I don't get nearly as much satisfaction from creating or maintaining plantings of stuff I can't eat. So, as my gardens and interest in gardening expanded, I focused on growing food. Beyond that, there are all kinds of details. I like lots of greens year-round. I don't like to spend lots of time in the kitchen processing stuff. I love to explore and figure out new methods. I love to breed things. I love to experiment, to try new things. I enjoy a certain amount of physical labor, but only so much of certain kinds. These "druthers"—my personal preferences—all influence the styles of gardening I do and the kinds of things I grow.

You may have completely different needs and desires from me. You may want, above all else, to plant a particular tree, bush, or vine whose aroma brings back wonderful childhood memories. Having that particular plant might be what it takes to make your house a home. You may want your garden to be primarily a source of cut flowers for your home. Or you may want a small garden or formal ornamental planting that you can control completely, unlike the rest of your life. So it will be just the right size and kind that you can keep utterly pristine, optimally manicured, and weed-free. Or you may want a yard and garden that produce the most satisfying vistas when viewed from the house.

When do you find yourself experiencing moments of intense joy in the garden? What kinds of gardening, what kinds of garden labor give you the most pleasure? Pay attention to yourself, to your feelings of inner satisfaction, joy, contentment, accomplishment, peace. In addition, note what kinds of plants, gardens, and gardening experiences give you the most frustrations or require the most unsatisfying labor. Honor your motivations and needs. Shape your choices of plants, gardens, and gardening styles accordingly.

Planning Versus Spontaneity

This year I interplanted trees with my spring-planted vegetables. Serious planning is obligatory when planting trees. Overwintering vegetables also take planning. It doesn't do to have the over-wintering vegetables spread all around the garden in such a way as to prevent fall or early-spring tilling of the entire garden. With one-season annual vegetable plantings there is more flexibility. I can plan the plantings ahead of time or I can just go out and start planting stuff, whatever and wherever I feel like at the moment. However, I make better use of space, water, and labor with at least some planning. In addition, often I'm doing trials of many different varieties of a given vegetable crop; these can require serious planning. Otherwise I find myself standing in the field juggling seed packets and planning while the cool hours of morning—the prime gardening hours—slip away. Whenever there is limited space, I can get the most food or information from it by serious planning. Even when I have plenty of land, I don't have plenty of time. I invariably want the most food and information for the least time and labor. This can best be achieved by serious planning.

I enjoy planning. However, I am not by nature obedient. I don't enjoy *following* plans. What I enjoy is spontaneity. Carol of yesterday might have carefully laid out exactly where to plant each of the twelve amaranth varieties in the amaranth trial and exactly where in the garden to plant them, but Carol of today, who is actually doing the planting, is likely to have other ideas. Doing Carol looks at the plans made and reviewed so carefully by Planning Carol and rebels. "Who the *#$! does she think she is ordering me around?" says Doing Carol. "What does *she* know? I'm sure I'm much smarter!" Doing Carol is bored with Planning Carol's plans. She thinks it would be more interesting to do something totally different. So she does. Sometimes Doing Carol does have some better ideas. But she often forgets some of the criteria that went into the plans. In addition, Doing Carol can get lazy and blow off eight of the twelve varieties.

Planning Carol has learned to deal with the rebelliousness and independence of Doing Carol. Planning Carol knows she can plan and design only up to a certain point. Doing Carol will respond positively to general, flexible plans with some options still in them. But if the plan is too elaborate or too rigid or too laborious, Doing Carol will trash the plan and do something else entirely. So Planning Carol creates "semi-plans" that spell out the basics, but leave room for flexibility. They especially spell out in detail certain aspects that are harder to figure out in the field. Planning Carol is unabashedly manipulative. She makes it easier for Doing Carol to follow the plans than to change them. The plans, for example, spell out the order of planting of the twelve different amaranths in a trial so that different colors alternate, making it easy to see where one variety stops and the next one starts. It's much easier to figure out that order sitting comfortably in an easy chair at home than standing in the field, so Planning Carol knows that Doing Carol will appreciate having that order already figured out and written down and will cooperate. Planning Carol also arranges the seed packets in order of planting and pre-labels the field markers (also put in order). This makes it much faster and easier for Doing Carol to arrange and plant the trial. It also makes it easier for Doing Carol to do the trial as specified than to change it. Planning Carol even arranges things so that those and no other varieties are what will be taken to the field that day. But Planning Carol doesn't specify exactly where in the field the trial will go, or how much space will be allotted to each variety. She doesn't make a map of the field showing what and how much of each variety is where. She doesn't even specify that the amaranth trial gets planted that day. Planning Carol just sends Doing Carol to the field with a box with the ordered seed packets and pre-labeled and ordered markers for planting the twelve-variety amaranth trial (and nothing else) and hopes for the best.

Doing Carol looks at the situation on the ground and skips the section that the tiller tilled only shallowly. She also skips the next potential bed spot because some of it is fine seedbed but some is lumpy. She chooses a different section for the amaranth trial bed, a spot with a fine uniform swath of seedbed that will be best for a bed with twelve different varieties that must be compared with one another. Doing Carol tosses some strips of plastic marker out to establish the outline of the bed and the right number of sections. Then she puts the pre-labeled variety markers down in order. (Markers are in order front to back, bound by a rubber band, representing the planting order left to right in the field.) Then Doing Carol looks at the amount of space in the bed, figures she wants twice as much of two varieties she already

knows will do well here, half as much space for the totally unproved varieties that might not grow here at all, and so on, and moves the section markers around a bit. Then Doing Carol records the bed sections and contents and approximate sizes in her field notes and sows all the seeds in all the sections. If Doing Carol is feeling too lazy to do such an elaborate trial, she can forget it and hoe out thistles this morning. Planting a trial with twelve varieties takes a certain amount of meticulousness. Lots of days Doing Carol just isn't up for meticulousness. She is best up for meticulousness first thing in the morning. So she can hoe thistles now or do hand watering, and come out tomorrow morning first thing and plant the amaranth trial. Doing Carol is unlikely to blow off eight varieties, or change the trial entirely once the order has been figured out, and the seed packets put in order, and the field markers also labeled and ordered. Instead, given the timing option, she will simply delay planting until she is in the right mood to deal with that kind of work. Doing Carol's need for independence and spontaneity is fulfilled by the choices that are left for her to make. And in fact, those choices are difficult to make except in the field. Back when Planning Carol tried to make those choices too, they usually had to be modified in the field anyway. Planning Carol and Doing Carol have learned to get along.

Structure, Labor, and Freedom

I most enjoy labor that doesn't have to be done at any particular time. I work in the garden most joyously when the work is optional. I might need to do it, but I don't need to do it that particular day. I'm doing it that particular day only because it's a good time for it and I feel like it. I like work best when I can quit anytime I want. I also relish being able to work on just one thing at a time for long periods of time. This means I like to avoid garden tasks that have to be done every day. I like to be able to not garden at all for a few days if I feel like it. Accordingly, I give plants enough space and use planting styles that allow me to water every few days or every week rather than every day.

Your preferences might be just the opposite. You might find, for example, that you really enjoy needing to hand-water your small garden each day. You might find it provides a structure and stability to your life that you relish. If so, you can go ahead and crowd your plants more than I do and get more production per unit area.

I prefer when possible to limit gardening to a couple of hours per day on the days I garden. Accordingly, I usually stagger my plantings so I don't have too much that needs to be weeded all at once. Planting too much at once leads to a need to water or weed too much at once. And for many crops, my labor at particular stages is the ultimate limit on how much I grow. I deliberately remember and consider how much work it takes to hill up potatoes, for example, when I am planning the size of my potato patch. For some storage crops such as dry beans, I use varieties with different maturity times to stagger the labor of harvesting.

When I was younger and less experienced, my lack of understanding of the labor implications of different gardening practices necessitated many garden marathons. If I planted too much at once, I could have too much needing to be weeded at once. However, I was using the gardening as a physical challenge and test when I was younger. I kind of didn't mind. And I wasn't breeding crops, growing seed crops for market, running a small seed company, or writing books. These days I prefer to avoid gardening marathons whenever possible. I like the labor spread out into manageable chunks. You may relish working in the garden for entire days at a time. On the other hand, you may need to limit your gardening labor to just a couple of hours every weekend. You will want to arrange your gardening style accordingly.

I have a weak back but powerful hands, arms, and legs. This affects what kinds of tools I can use most easily and the gardening styles I choose. In addition, my home garden is large, up to about half an acre (0.2 hectare) that I tend by myself these days completely with hand tools other than the yearly visit from the tractor guy. (This doesn't count the co-cropping of seed crops that I participate in growing but don't have to weed or water.) That size garden is manageable only if I garden efficiently, and in a way that is as easy on my body as possible.

I prefer direct seeding instead of transplanting wherever possible because, with the right methods, direct seeding can be done standing comfortably upright instead of squatting or kneeling or on hands and knees. And I use the special kinds of long-handled hoes and the counterintuitive grips for using them that allow me to work with a straight back instead of needing to bend over as with most hoes and styles of hoeing. I also use planting patterns that allow weeding by hoeing as much as possible rather than requiring me to bend over or get down on my knees to hand-pull weeds. Your bodily strengths and weaknesses, thus your needs and preferences, might be quite different. For the labor implications of various styles of gardening and the choice of particular tools see chapter 5, Labor and Exercise, in my previous book, *The Resilient Gardener*.

道

Live wood is soft and tender. Dead wood is hard and strong.
The live leaf is supple. The dead leaf is brittle.
The army that isn't flexible never wins a battle.
The tree that survives the storm is the one that bends in the wind.

Yield and prevail. If you can bend, you can be preserved unbroken.
If you can flex, you can be kept straight. If you can be emptied, you can be filled.
If you can be worn out you can be renewed. Those with little can
receive much. Those with much can be confused.

Know the male but hold to the female. Know the white but hold
to the black. Know the peaks but hold to the valleys. Know complexity
but hold to simplicity. Know sound but hold to silence.
Know action but hold to stillness. Know knowledge but hold to innocence.
Know the manifestations but hold to the mystery.

The Tao is like water. Water takes the lowest way.
It flows around obstacles. It has no projections, but it penetrates that
which has no crevices. It's the softest of all things, but it grinds rock.
It resists nothing, but is irresistible. It asks for nothing, but nourishes
everything. It strives for nothing, but transforms everything.

道

CHAPTER FOUR

Flexibility

*Choosing Gardening Styles and Methods. Getting the Most
from the Small Garden. Volunteers. How to Eat a Weed—
Dandelions, Lambsquarters, Purslane. The Prepper's Garden.*

One afternoon in ancient China the Taoist sage Chuang Tzu and a disciple
were on their way to dinner at the home of the district magistrate. It was
an era in which every change of government resulted in the execution of
virtually everyone in any position of power, and governments changed often.
Chuang Tzu avoided politics and positions of power. On the way to the
district magistrate's, as they passed under a particularly large old shu tree, the
student said, "I have heard that the shu tree lasts so long by being useless to
man. It produces no fruit or nuts so is not pruned and controlled like orchard
trees. And the wood is too crooked and gnarly to build with. And it produces
little heat when burned so is not even good for firewood. So shu trees survive
and thrive while more useful trees are cut down or controlled. It is well
known that you follow the model of the shu tree, and make yourself useless
to the state by refusing to serve in the government. Isn't that so?"

"True," Chuang Tzu responded.

"So why then are we going for dinner with the magistrate?" the student
asked.

"He invited us."

The student wasn't at all satisfied with this answer, but they had
reached the courtyard of the home of the magistrate, so he dropped the
subject. As they approached the house, a cook came out and yelled at a
boy, "Hey, Boy! Fetch me a hen to butcher for dinner."

"Which one do you want?" called the boy. "The one that cackles or the
one that doesn't?"

Now, Chuang Tzu's disciple knew that a hen cackles right after it lays
an egg. So a hen that doesn't cackle is a hen that doesn't lay. The disciple
stopped and listened to see which hen would be killed—the useful or the
useless one.

"Fetch the one that doesn't cackle, of course," yelled the cook. "Why should we be wasting good grain and scraps on a hen that doesn't lay!" The boy vanished to do the cook's bidding.

"Master," said the student to Chuang Tzu as they entered the hall. "I don't understand. The tree lives long by being useless. But the hen dies by being useless."

"No one principle is good for all situations," Chuang Tzu responded.

Choosing Gardening Styles and Methods

About a decade or so ago I hosted a team of writers and photographers who came to visit my gardens. In that era I had gardens in three different spots, and we went to all of them. First I showed the team my small backyard garden. It consisted of eight permanent raised beds with wooden sides on a gently sloped hillside. I had inherited the raised beds. However, beds of some kind were the only sensible way to garden in that yard. The bits of usable ground were too small and broken up by dense shade from the house or trees on neighboring land or septic drainage fields to work with a rototiller. In addition, the land was sloped and would be vulnerable to water erosion if tilled. It was also very heavy clay and the soil was shallow. Bringing soil in and putting it on top of the shallow clay to make raised beds, as the former owner had obviously done, was a sensible solution. The beds ran roughly north and south. This also made sense, as it paralleled the slope of the land. Sometimes my planting rows ran along the beds and sometimes across them, depending on the plants and how much space they needed and how I expected to do the weeding. Tomatoes need a lot of space, so I ran a single row down the

middle of one bed. My small fava bean patch and my garlic patch I ran in short rows across each bed to facilitate weeding with a hoe. It's easier to hoe rows that run across a bed than to try to reach over one or more rows to hoe rows running parallel to the access paths.

Had I created the beds myself, I would not have made the wooden sides for them. Unless the slope is very steep, wooden sides just get in the way and increase the work. The sides interfere with using a hoe either within the bed near the sides or outside the bed to maintain the path. With gently sloping beds with no sides, I could have mowed the paths. As it was, the beds were too close together for a mower, so the center of the paths had to be hoed and the edges around the wooden frames hand-weeded. With the sides, it was also harder to turn the soil over with a shovel within a bed. And alas, those frames were built to last. I did manage to dismantle some of them ultimately, but it took hours of work per bed.

The beds measured about 3.5 feet by 12 feet, or 42 square feet each (1.1 m by 3.7 m, or 4.1 square meters each). With eight beds I had a total of 336 square feet (33 square meters) for my home garden. This is where I grew my tomatoes, greens, garlic, herbs, summer squash, green onions, green beans, peas, cucumbers, and anything else that needed to be tended or harvested regularly. In a friend's field a ten-minute drive away I had a huge winter squash patch. I tended and watered it about once per week, but only needed to harvest it at the end of the season. My third garden was on a farm about an hour's drive away. There I had dry corn for grinding, dry beans, and potatoes. On that farm I designed the plantings and participated in the planting and harvesting, but had a cooperative arrangement so that I didn't have to do the irrigation or weeding. So except for the planting and harvesting, I just visited that garden once or twice during the growing season. The corn and beans were irrigated. The potatoes were grown without irrigation.

In my home garden, I maintained all the beds by turning the soil over with a shovel between crops. I gardened mostly year-round. Tilling tends to force you into planting everything all at once and not having ground ready for planting at times that might be optimal. Hand-dug garden beds can be prepared and planted as you need them throughout the year. You can even cover a bed with plastic to keep the rain off so you can work it when you want instead of needing to wait for the right conditions to occur naturally. The first planting of peas and fava beans can usually be made here in February or March, but not if you are dependent on tilling. You are not likely to be able to till that early. With hand-dug beds, I was doing a little digging and planting nearly every month. It was very easy to garden year-round.

I watered the beds with a hose by hand, giving quite different amounts of water to each bed or even to individual plants in each bed. The garlic patch I didn't water at all. I planted it in mid-October, and it grew on winter rains. In late June and July when the groundwater dries up, the plants dry out naturally and optimally. I harvested the bulbs in late July. A hard spray aimed directly at the kale leaves washes the aphids right off them. In a bigger garden that is watered with sprinklers, you can't do nearly as good a job of providing each plant with just what it needs as you can in the small hand-watered garden.

I interplanted in most beds. This wasn't because of any presumptions as to what the plants preferred but because, when space is so limited, you usually replant any space as soon as it becomes available. You can't afford to waste garden space by delaying replanting until the entire bed becomes available.

The tomato bed had a light seasonal mulch, but the rest of the beds were unmulched. With the tomatoes, I needed to ameliorate changes in moisture levels between waterings so that the fruits didn't split when watered.

I practiced vertical gardening wherever possible. That is, I grew the big, vining indeterminate forms of tomatoes instead of determinate kinds, and I supported them with sturdy cylinders about 2.5 feet (0.8 m) across and 7 feet (2.1 m) high made from concrete reinforcing wire. Such large-vined caged tomatoes take only a little more space than determinate bushy types, but produce much more fruit. I also grew pole beans rather than bush beans, using another concrete reinforcing wire ring with a double row of beans planted around it. Pole beans produce about twice as much per unit of row as bush beans. In addition, many pole bean varieties have much bigger beans with better flavor than most bush beans. Other rings only half as high held edible-podded peas in wide rows arranged in circles around the supports. Others held cucumber vines.

Vertically grown crops cast shade. I broadcast my greens mostly in small patches wherever they fit, often in the shadier spots created by the vertical crops. Greens generally do fine in partial shade. And having lots of little patches of greens planted here and there at different times spreads out the harvest.

I always plant tomatoes as transplants, as there is no real choice given their tropical nature and my temperate climate. Given my druthers, I prefer to direct-seed cukes and squash, but in this situation I usually used transplants. With such a small garden, planting transplants is a good way to get the most out of each bit of soil. The transplants go into the garden only when they are big enough to take full advantage of the space. I direct-seeded the peas, beans, and greens.

After we toured my tiny home garden, the visiting team and I jumped in a big van and drove to the squash patch I was growing on land that belonged to a friend about a ten-minute drive away. My field was tilled along with my friend's gardening area by hiring the tractor guy each spring. That tilling pattern didn't fit with year-round gardening. But it was fine for growing warm-season crops that are planted late enough in spring so that the ground has dried out enough for the tilling. After that initial hired-out tilling, I did all the weeding with hoes and hand tools.

I direct-seeded all the squash in rows that ran east and west. This had nothing to do with the direction of the sun, though. The configuration of rows was perpendicular to the access road and the water line. With this orientation I could easily get a hose between the rows to facilitate watering. If I set the rows up any other way, I would have had to drag sprinklers and hose over the rows, disturbing and smashing plants in the process.

Finally we loaded everyone into the van and drove for an hour to a farm belonging to a friend of mine where I had plantings of dry beans of several kinds, as well as flint corn, potatoes, and some other staple vegetables. I planted and harvested these in cooperation with my friend, who did the weeding and watering. This friend tilled the fields with a tractor in spring and used a rototiller between the rows during the season. Everything was planted in rows. This made sense given the kinds of crops and my friend's pattern of using a rototiller for cultivation between rows. The orientation of the rows was parallel to the slope to prevent erosion. Most of the plantings were watered with sprinklers; one part was on drip lines. The potato patch was unirrigated.

After looking at all three gardens, the visitors and I sat in the shade of trees bordering a small river to eat our lunches, and they presented me with a written list of questions, several pages' worth. Did I garden in rows or in beds? Did I direct-seed or use transplants? Did I irrigate or not? Did I mulch or not? Did I use a rototiller or hand tools? How did I run my rows with respect to the orientation of the sun? North and south? Or east and west? And so on.

I read through the questions, then handed the list back.

"Yes!" I answered cheerfully.

Getting the Most from the Small Garden

For maximum yield of food for the amount of space, use some variant of beds rather than rows. Rows are better for certain crops such as corn and winter squash. But these aren't crops you usually grow in the small garden where space is at a premium. Beds may be permanent or just for the season. They can be raised or level with the rest of the garden. Raised beds dry out and warm up faster in spring but are harder to keep watered in summer. Permanent raised beds are usually dug and worked by hand. Temporary beds might be dug by hand or formed with a tiller. If you use a tiller on your garden annually, you then just lay out your plantings in beds instead of in rows. That's all there is to it.

In the small garden you usually plant more intensively than in the larger garden. That is, you give the plants less room per plant. You can get away with this only by making your soil more fertile than is typical for the larger garden. However, this is easy to do when you have a small garden. It also helps if the fertile, soft soil layer is deeper than is the case in the typical larger tilled or rototilled garden. For absolutely maximum yields, use permanent beds and double-dig them. The classic book on this subject is by John Jeavons and is titled *How to Grow More Vegetables (and Fruits, Nuts, Berries, Grains, and Other Crops) than You Ever Thought Possible on Less Land than You Can Imagine*, 8th edition. For anyone interested in getting the most food for the least space, whether in double-dug beds or not, this book is the essential reference. It has more hard numbers than any other garden book I know: amounts of time it takes for various tasks; yields per garden bed for different crops; instructions for growing everything. The book is usually referred to simply as *How to Grow More Vegetables*. Be aware that there is now another book by that title. Make sure you get the book by John Jeavons.

Crowded plants need more water and usually more frequent watering than plants that have more space. Part of why larger gardens are often planted mostly in rows is that in many areas of the country generously spaced plants in rows don't need irrigation. In the maritime Northwest, most of the warm-season annual crops require irrigation even when grown in rows with generous spacing. They simply don't need to be watered as often. The intensively planted bed may need to be watered every day during hot periods and every other day during most of the summer. The less intensively planted bed may need to be watered only half as frequently.

Compost is usually a big part of the picture for a smaller garden. You can compost all your table and garden wastes and grass clippings and leaves and use the compost to enrich your garden. That is, you take the fertility mined out of the soil by trees and grass on a much larger space and concentrate that on your small garden. Leaves are usually available from neighbors. (Leaves don't have enough nitrogen to compost by themselves. But urine has lots of nitrogen . . .)

Use vertical gardening wherever possible as described in the previous section.

For more efficient use of garden space, use transplants wherever possible.

You will inevitably multicrop and overlap crop plantings in various complex ways, working out your own patterns for what works and what doesn't. Most of what has been written about "companion planting" seems to be just made-up wishful thinking rather than anything that has been seriously tested in any way. Just try things, whether they are the recommended combinations or not.

Don't try to grow everything. Focus on the plants that give you the most delight. It's great to be a grown-up. If you don't like brussels sprouts, you don't have to eat them, and you don't have to grow them either. If you love tomatoes, grow lots of them.

Also focus on the crops that provide the greatest nutritional value.

Greens, especially cooking greens (not lettuce), are the plants that are likely to make the biggest nutritional contribution to your life. Not all fruits and vegetables are created equal. You don't actually need yellow vegetables or even fruit if you have plenty of greens. Greens can provide the nutrients that other vegetables provide as well as lots more that are available only from greens. Stem vegetables usually are less nutritious than bud vegetables. And loose leafy greens are the most nutritiously valuable of all—lettuce excepted. Plant lettuce only to the extent that you must. It is one of the least nutritious of greens. Spinach is much more nutritionally valuable, but there are many greens that are equally or more nutritionally valuable than spinach that can be grown more easily and yield better and can be used in all the same ways.

Plant largely the eat-all greens I describe in the chapter on the subject, rather than green vegetables that provide just a single head for the space they occupy and that take all season to produce even that. Heads are needed for storage or shipping, so they are very valuable commercial crops. You can produce much larger harvests of leafy eat-all greens in far less time on the same amount of space as is required for a few heads. Furthermore, with the eat-alls you can get several crops per season. Growing and freezing all your family's winter supply of greens from even a small garden is possible with the eat-all approach.

Greens are generally highly shade-tolerant. If your garden is small because most of your property is shaded, you can make some extra beds in the shade just for greens.

Learn to make superb soups and stews and make them a regular part of your diet. Many of the best and easiest-to-grow garden plants are best used in soups and stews. And much in the garden tends to go to waste if you don't make soups and stews.

Fill out your salads, batches of cooking greens, and soups and stews by using various edible parts other than those usually used. Carrot tops make good salad or cooking greens. When you harvest the carrots, cut off the tops and use them immediately or put them in a plastic bag to refrigerate separately from the carrots. If you leave the tops on the carrots they transpire away water, the greens wilt, and the carrots go limp.

Cabbage and broccoli leaves are edible when young and tender. Young leaves of horseradish, sliced fine, can jazz up a salad. The bolting scapes of kale, cabbage, and other brassicas are edible. Runner bean flowers are edible. Nasturtium leaves make spicy, fiery greens that are great in salads or sandwiches, and the flowers are edible. Young turnip leaves as well as radish leaves are edible. (As with carrots, cut any greens away from the roots and use them promptly or refrigerate them in plastic bags.) Beet leaves are eaten and enjoyed by many people. And of course, don't waste any edible weed.

Make use of container gardening where appropriate. You may have only a tiny patch of soil that's workable for a garden, but have more room on the corner of the driveway, or on a patio, a porch or deck, or even on the roof. You can grow a lot of food in containers if you select the right varieties. Most books on container gardening focus on flowers or ornamentals. A great book on growing food in containers is *The Bountiful Container: Create Container Gardens of Vegetables, Herbs, Fruits, and Edible Flowers* by Rose Marie Nichols McGee and Maggie Stuckey. Rose Marie Nichols

McGee is co-owner of Nichols Garden Nursery, so Nichols is the first place to look for the varieties suitable for container growing.

When I lived in town in Corvallis, some of my gardening was done in beds under trees in the strip between the sidewalk and the street. That was almost the only space available. There was plenty of sun. However, the roots of the trees were aggressive. Each summer they grew up into the soft, irrigated garden soil in the raised beds and started stunting the vegetable plants. By the following spring, the raised beds were so heavily occupied by tree roots that it was almost impossible to turn over the soil, and unless I cut the roots back I wasn't able to grow vegetables even early in the season. Digging and chopping out all the roots was tremendously laborious. Faced with that situation again I would construct large wooden boxes the size of the beds, prop them up on concrete blocks to keep an air barrier between the tree roots and the boxes, fill them with garden soil, and garden there. I once saw such bed-sized containers at the home of a friend. Those containers had been constructed for a different reason. The gardener had made and planted them prior to a move so as to be able to bring some garden plants along and not lose an entire gardening season. Containers of many kinds can be part of the small garden. They can let you garden in places where you don't have soil suitable for digging or tilling.

Finally, if you plant winter cover crops, plant edible ones. Austrian field peas, daikon radish, and mustard all produce edible greens. You can also use vigorous overwintering varieties of culinary peas, leaf-bred radishes, kale, and mustards as cover crops. For the field peas, the top 3 inches (7.6 cm) of the growing shoot along with its leaves are edible throughout the winter and spring.

Fava beans are a common winter cover crop. The tops of fava bean plants with all their attached young leaves are edible unless you have a certain enzyme deficiency (glucose-6-phosphate dehydrogenase), in which case they can be dangerous or even lethal. For this reason I don't recommend fava greens or fava green beans as edibles.

Volunteers

Early last spring I planted something or other in the northeast corner of my garden. I don't quite remember what it was because it never saw fit to germinate. Instead what I got was radishes coming up all over the area. These were leafy radishes, that is, mostly radishes bred for their edible leaves instead of for their roots. I had leaf radishes in that area the year before that went to seed and got tilled in. Whatever I had planted was probably also an early-spring green of some sort. Apparently I thought I wanted it more than extra leaf radishes. I already had another patch of leaf radishes I had planted on purpose. I could hoe the volunteer radishes down and replant with something else. But the radishes were already there and growing fast . . .

There are lots of things you can do with extra leaf radishes. There are all the ways to use them fresh that I discuss in the chapter on the eat-all greens garden. In addition, you can make kimchi from them, or blanch and freeze the greens, or dry them for adding to winter soups and stews, or to make a lovely herbal tea. I had thought I wanted some other kind of patch, but ended up with an extra radish patch. So be it.

Many garden crops, if allowed to go to seed, will self-sow, then come up the following year as volunteers. Sometimes letting that happen is the easiest way to grow the crop. I have a patch of 'Alexanders Greens' (*Smyrnium olusatrum*) that has been going for six years with no tilling, tending, weeding, or watering of any kind. It is in such a shady place that even grass won't grow. (See the photo section.) The plants germinate with the fall rains and grow into stalks with three leaflet leaves that have a flavor like a cross between celery and parsley. The young leaves and stalks and supposedly even the roots are edible. (I haven't tried the roots, as my patch is in solid undiggable clay.) The young leaves and stalks are prime and available all winter and early spring. I use them both in salads and as cooked greens. In late spring the plants bolt and produce seed. This life cycle fits perfectly into our patterns of available rain, so the plants need no irrigation. In the maritime Northwest, 'Alexanders' provides a superb no-care supply of tasty greens in winter just when fresh greens are most appreciated.

Hardy kales that are capable of overwintering also frequently volunteer. I used to just weed around them and use them while they were young. These days I encourage their volunteering by sowing the seed in just the right places for a more serious volunteer kale harvest. Call it semi-volunteering. I spread the seed but don't give any further care. The trick is to spread the seed in July or August in places that will be irrigated during the rest of the summer but won't be tilled that fall. These sowings will produce plants that provide food all winter.

'Magentaspreen Lambsquarters' (*Chenopodium giganteum*) is a type of greens I often plant. Also known as giant goosefoot, it is a relative of lambs-

quarters and quinoa that has bigger leaves and more succulent stalks and is slower to bolt, thus producing greens for a much larger part of the season than regular lambsquarters or quinoa. The growing tips of 'Magentaspreen' have a magenta blush to them that is distinctive. Plants can grow to 8 feet (2.4 m) high, producing succulent growing tips and big edible leaves most of the summer. But 'Magentaspreen' is unruly. It's one of the best of all greens, though, so I give it some slack.

Here's how I usually grow 'Magentaspreen'. I sow the seeds in a patch in early spring. Something else comes up, weeds or volunteers, never the 'Magentaspreen'. After I have given up on the 'Magentaspreen' and hoed down the weeds and planted something else, just a few 'Magentaspreen' seedlings come up as volunteers. Then the 'Magentaspreen' volunteers in that area for a number of years. Meanwhile, 'Magentaspreen' volunteers cheerfully and makes a dense lovely stand in some other patch where I tried to grow it unsuccessfully three years earlier. So I pretend I planted that patch and just water and harvest those. I plant a patch of 'Magentaspreen'. I get a patch of 'Magentaspreen'. Just not quite where and when I intended. Planting later in the year doesn't change the pattern. The seed sits there for quite a while, then finally a small portion of it germinates after a month or two. The rest remains in the ground to become volunteers in future years. These days I just toss 'Magentaspreen' seed around where I think I might want some next year, and plant other things on top of it, trusting the 'Magentaspreen' to mostly not come up the first year.

I have taken to treating 'Magentaspreen' more as an edible weed than a garden plant. It behaves like a weed, but it produces as much or more edible food for the space or care as any domesticated greens. I do sometimes sow it, but I mostly grow it by tolerating and encouraging its volunteering.

'Magentaspreen' behaves like a wild plant for a very legitimate reason. It *is* a wild plant. That is, it was until a few decades ago when my friend, the seedsman and plant breeder Alan Kapuler, brought it in from the wild, pronounced it prime eating, named it, and began treating it like a garden plant. (A. K. coined the word *spreen* for spring greens, the succulent young shoots and leaves of plants that make such great edible greens in spring.) 'Magentaspreen' still has all its wild-plant seed dormancy mechanisms. Part of domestication usually involves eliminating those seed dormancy mechanisms. This makes the seed dependent on us to plant it at the right time instead of independently deciding for itself when to germinate. 'Magentaspreen' has yet to learn to depend on us, trust us, and grow where and when we tell it to. The 'Magentaspreen'–human relationship is a work in progress.

(Note: I don't save seed from the 'Magentaspreen' volunteers. If I did I would be selecting for the seed dormancy mechanisms and encouraging the plant to evolve in the direction of being even less controllable. Instead I am selecting seed from plants that germinate in the first year they are sown, hoping to select ultimately for a line of 'Magentaspreen' without the seed dormancy mechanisms, a line that is more controllable and better behaved—that is, a plant that is truly domesticated.)

Volunteer potatoes don't find me very friendly or tolerant. They can be a major source of late blight and other diseases. Their tubers have not been subjected to the kinds of screening and culling for disease I do with the tubers I save and plant as seed. I dig up and destroy all volunteer potatoes. In addition, volunteers really interfere with crop rotation, which is especially critical for crops in the Nightshade family (Solanaceae). As Chuang Tzu noted, no one principle is good for all situations.

How to Eat a Weed— Dandelions, Lambsquarters, Purslane

Edible weeds are the quintessential volunteers. We call them weeds partly because of their seed dormancy mechanisms. Some of them produce food that is every bit as tasty and nutritious as our domesticated alternatives. However, they don't produce as much or over as long a period as the plants we sow on purpose. When edible weeds volunteer in our gardens, we are often a bit conflicted. Our inner gardener resents these plants that are interfering with our desire to grow something else and worries about their going to seed. Simultaneously, our inner hunter-gatherer shouts gleefully, "Free food without all the grubbing around! Leave them be! Just wait a bit and harvest them!" My favorite edible weeds are dandelions, lambsquarters, and purslane.

Dandelions are perennials. I weed them out of my vegetable garden, where their habit of making seed regularly all season is too troublesome. But I encourage dandies in my lawn. The seed can only occasionally germinate and make a new plant in established sod. And as far as I'm concerned, every lawn could only be improved by having more dandelions. If I could, I would get rid of all the grass and have a lawn that was just dandelions.

The best place to encourage dandelions is a spot with partial to nearly full shade. Dandelions can grow in areas so heavily shaded that grass doesn't grow or grows only poorly. In these shady spots the dandelion leaves grow tall, are mostly large and undivided, and stay tender and non-bitter much longer in spring. The leaves remain good to use even after the plants start flowering. By contrast, dandelion plants growing in full sun have smaller, narrower, highly divided leaves that become bitter much earlier, and the leaves are usually too bitter to use after the plants start flowering. In the maritime Northwest the plants are best grown in areas that are not irrigated. If irrigated they get powdery mildew in summer. In our region dandelions grow mostly in winter and spring on natural rains and go dormant or grow only very slowly in late summer. The most luxurious stand of dandies I ever saw was a solid stand of unirrigated plants about 2 feet (0.6 m) tall covering an abandoned compost heap in the shade north of a barn.

People frequently talk about boiling dandelion leaves in one or more changes of water to remove the bitterness. I prefer to use the full-sized but young leaves before they have much bitterness, as well as mixing them with other greens to dilute any remnant bitterness. Boiling greens in changes of water is laborious. In addition, lots of valuable nutrients are discarded with the water. I mention the boiling in changes of water only in case you are ever in a situation where you have no other greens and the lack is desperate enough that you would want to use older greens. I always prefer to drop greens into soups and stews and cook them just briefly before serving, thus retaining all the nutrients that leach out in boiling, or to cook the greens briefly in water that is also saved and used.

Lambsquarters (*Chenopodium alba*) make an excellent cooking green from late spring to early summer. When the plants are about 6 to 8 inches (15 to 20 cm) high, the top 3 inches (7.6 cm) of stalk with leaves is edible and can be snapped off each plant. (Some of the leaves lower than that are also tender and can be harvested too.) The plant then branches and produces another harvest or two of edible tips with leaves before bolting. The lower leaves and stalk are too coarse or woody to eat. I often hoe around the plants in early spring and take a harvest or two of greens from them, then hoe them down in midsummer before they go to seed. Lambsquarters that emerge in midsummer go to seed so early and fast they produce little food; I usually try to eliminate all the lambsquarters starting at midsummer. There is actually a lower section of garden with poorer soil that I till but don't usually plant or tend. I let the lambsquarters and 'Magentaspreen' go to seed and volunteer there. All the crops I describe in the eat-all greens garden chapter produce much more food for the space that is easier to harvest than lambsquarters. But the lambsquarters is free.

Some people use lambsquarters in salads. I find them too dry and fuzzy for salads. But they make great cooking greens. To use lambsquarters as cooking greens, I cook them a few minutes in boiling water, drain them (saving the water for soups, stews, or tea), then dress the cooked greens in the various ways I describe in the eat-all greens garden chapter. Or I drop the raw sliced greens into soups and stews during the last few minutes of cooking.

Lambsquarters are so dry they actually absorb fluid as they cook. I sometimes take advantage of this characteristic to make thick dips and sauces. For example, I might start with a can of Progresso Clam Chowder in an 8-inch (20 cm) Pyrex bowl, toss in a can of albacore tuna and a good bit of curry powder and plenty of cheese, and zap the bowl in the microwave oven for a few minutes to heat everything and melt the cheese while I'm slicing up the lambsquarters. Then I add the lambsquarters and zap about three minutes more until the greens are cooked. The lambsquarters absorb so much water that the soup then becomes a rich cheese-tuna-clam-lambsquarters sauce I can put over rice or polenta or use as a dip.

One of my favorite things to do with lambs-quarters is to dry it. The stems that are succulent before drying become unpalatable little sticks after drying. So I don't chop up the succulent tips with the stems. Instead I dry the tops of the plants whole (the entire part of the top that has edible leaves, the lower part of which has a woody inedible stem). Then when the plants are dry I hold each plant by the stem with one hand and use the other hand to strip all the dry leafy material off in one motion, leaving all the stems and even most of the leaf veins on the stalk. I end up with a pile of small pieces of dry leaf that has a marvelous flavor in winter soups and stews. (You just drop it into the soup or stew and boil it a couple of minutes.) The dry leaves absorb moisture from the air very quickly, so they should be packaged in small airtight containers. The dry leaves make even more delicious soups, stews, or sauces and dips than the fresh ones do.

Dried lambsquarters leaves also make one of the most delicious herbal teas.

Purslane often volunteers in corn patches. It is a succulent plant with thick mucilaginous leaves and stems. It has very high levels of omega-3 fatty acids. There are garden forms that are more erect than the wild forms, whose stems hug the ground. In my garden, by the time purslane stems are much bigger than about 6 inches (15 cm) long the thick leaves become infested with worms or grubs of some sort that burrow around and eat the insides of the leaves. Every leaf becomes full of such worms. This happens every year. So I leave my purslane volunteers until they are about 4 inches (10 cm) in length, then harvest the entire plant for the soup or stew pot before the worms appear.

The Prepper's Garden

During World War II the existing food-producing capacity of the United States was strained by the need to feed the Allied armies as well as much of the remaining free world. Simultaneously, the labor available to work farms dropped drastically as men joined the military. As part of the war effort the Department of Agriculture mounted a massive campaign to encourage all Americans with any access to land, from urbanites and suburbanites to rural people, to grow gardens—"Victory Gardens." More produce from gardens would reduce domestic demand and free up more of the farm produce for the war effort. The nation responded enthusiastically, encouraged not just by patriotism but also by domestic food shortages and food rationing. The result was an estimated ten million new urban and suburban gardens in addition to five or six million farm gardens. (See chapter 26 in *A Call to Arms: Mobilizing America* for World War II by Maury Klein.) The transition from non-gardener to successful Victory Gardener was not necessarily smooth, however. Seed companies reported receiving a flurry of orders from new gardeners, among them requests for seeds of coffee and succotash.

Gardening is a survival skill. It can help you through crises of many sorts and sizes ranging from simply being without a job or short on cash for a while to serious mega-disasters—wars, famines, natural disasters, and social upheaval of many kinds. However, a gardening book and a can of "survival seeds" do not a successful gardener make. That can of survival seeds is not likely to contain varieties optimal for your area. Many might not even grow in your region at all, even if you knew how to plant and tend them, which you don't if you haven't gardened before. And unless you grow long-storing staple crops and know how to store or put up produce, even a successful summer garden would leave you very hungry in winter.

If you are new to gardening and coming from the perspective of the prepper, your first job is just to learn to garden in your region.

Second, you should learn to save seeds and build your own emergency long-term seed bank of varieties you like and know how to grow. (For more on this, see the final chapter of this book.) Seeds are the first and most critical input you would be unable to get in the event of most disruptions, large or small. Third, you need to learn how to grow and preserve food for the winter months. Fourth, you should do some experimenting with hard times gardening, that is, gardening dependent on only the resources you would be likely to have in hard times. This might mean, for example, no imported fertilizer, no bought seed,

and no electricity (which might mean no irrigation). Fifth, you need to grow not just salads and summer vegetables, but also staple crops such as potatoes, grains, and beans that provide serious amounts of calories and protein and that can be stored for winter. I discuss all these subjects at length in *The Resilient Gardener: Food Production and Self-Reliance in Uncertain Times*. That book has major chapters, mini books actually, on growing, storing, and using each of the staple crops that are most practical for most gardeners—potatoes, corn, beans, squash, and eggs. I also recommend my friend Steve Solomon's book *Gardening When It Counts: Growing Food in Hard Times*.

What if you don't have the land or just aren't at the right place in your life to grow a big garden? Fortunately, we humans have been actively specializing and trading for at least the last fifty thousand years and are very good at it. The modern humans who invaded Neanderthal Europe carried tools made from rock traded over long distances. The Neanderthals, by comparison, had tools made of local rock. I rather suspect it was trading that made humans human. We swapped our ideas and innovations along with the physical stuff we traded, I believe, creating a kind of recombination of ideas, inventions, and culture that led to much faster cultural elaboration than has occurred for any other animal. I think that, for us, the drive to trade is actually built right into our genes. Speculations aside, if you look at historical accounts of disasters of many sorts that disrupted the established trade relationships, you find people still trading furiously and reestablishing trade relationships almost instantly.

This trading may take many forms. It may not be called trade. The exchanges may not be simultaneous in time or reciprocal. Many people volunteer to help me plant and harvest my seed crops because they appreciate my breeding and releasing public domain varieties and realize it doesn't pay for itself. A friend of mine stores a trailer for another friend, whom we want to help however we can because she has helped to build 652 Victory Gardens in the last four years for people who would not have been able to garden otherwise. We are not the ones who received the Victory Gardens. But we want to honor her contributions to the tribe and support her by doing whatever we can when she needs something. That's how it works. Sometimes it is "I'll trade so much of this for so much of that." But often you simply make gifts of some of the best you have to those you care about, and they do likewise.

Anthropologists tell us that credit has existed in every culture since long before the invention of money. (See *Debt: The First 5,000 Years* by anthropologist David Graeber.) Contrary to the assumptions of economists, credit comes first; money is a more recent invention. Defining trade in terms of money rather than credit actually changes the nature of the trade transactions and has liabilities. It takes the relationships out of the transactions. But the more ancient, normal, natural, and perhaps most emotionally fulfilling pattern for us is to engage in money-less but relationship-rich gifting and trading. This is still the kind of trading most of us do informally when left to our own devices and when dealing with people we know and care about. And no monetary systems or bankers need apply. The essential part is simply that you provide things you are capable of providing to others, and various others provide you with things they are better at providing. So if you have only a small garden and can't grow all your food

in good times, let alone bad ones, never mind. For more than fifty thousand years individual humans have not needed to provide everything for themselves. You don't need to either.

If you are a prepper at heart, without the ability to grow a big garden at the moment, here are some other approaches. First, I suggest learning to store a year's supply of grains and beans and working them into your ordinary diet so that you are turning over your supplies regularly. If you have much processed food in your diet and replace it with such staples, this will improve your diet and lower your food costs as well as increasing your family's food resilience. It's a better food-buying and -use pattern for good times as well as a preparation for potential emergencies.

Second, I suggest you develop some food-handling skills that can be traded for other people's food-growing skills. If, for example, you become an expert at making sauerkraut or kimchi or other fermented vegetables, you can probably get someone to grow the vegetables for you in exchange for a share of your pickled vegetables. Then there is the all-time classic, the lady who can make the best apple pies in the neighborhood. She never has to do without pies, even if she has no fruit trees. Anyone will give her enough fruit for a couple of pies if she gives them one. If you learn to can fruit you can usually get the fruit for free in exchange for a share of the canned fruit. If you live close to me and want to can tomatoes, I'll be glad to grow and pick all the tomatoes in exchange for half the processed products. And I would happily trade great polenta corn, cornbread corn, dry beans, or the world's best winter squash for canned fruit. Just ask me.

Learning how to store or use a staple crop is often more difficult than learning how to grow the crop. It took me lots longer to learn how to store potatoes optimally under my conditions than to learn to grow them. You can buy potatoes in bulk from local farmers and start experimenting with storing them. If, for example, you determine that putting them in paper bags on shelves in your attached garage (as I do) will keep certain varieties for up to eight months, and become familiar with the storage lives of different varieties, then you can start buying potatoes in bulk in fall from gardening friends or local market farmers. If something happens and it turns out to really matter that you have a generous stockpile of potatoes that winter, it is not going to make any difference at all who grew them.

You may be able to swap work in other people's gardens in exchange for crops. If you have neighbors with an untended orchard where the fruit is going to waste, make a deal with them to tend the orchard and do the pruning and harvesting for most of the crop. Or volunteer to help out in a neighbor's garden or on a local farm in exchange for crops. Growing major crops of anything produces labor bottlenecks. Offer to help out during the bottlenecks in exchange for some of the food. With corn, dry beans, and squash,

planting and harvesting are the labor bottlenecks. Planting, hilling up, and harvesting are the labor bottlenecks for potatoes, especially the harvesting. I have sometimes swapped a share of the potatoes to people in exchange for help digging them up.

We humans are social creatures. We don't thrive in isolation. We don't need to produce all our own food, tools, housing, recreation, or protection. Instead we use our individual brains, skills, talents, and resources to specialize. Then we trade the best products of our brains and labors for those of others. We aim not for independence but for honorable and companionable interdependence. We don't need to have all possible abilities and skills. But we do need to have some skills that give us something to contribute. In good times, our jobs are a big part of what we contribute. In hard times, our ordinary jobs may become irrelevant. Simpler skills, including knowing how to produce, preserve, and use food may be what matters. Having such skills enhances our resilience. A community in which many people have such skills is a more resilient community, a community best positioned to thrive in good times and survive the rest.

道

Too much light causes blindness. Too much sound causes deafness.
Too much indulgence in physical pleasures causes distraction and dissipation.
Having too many possessions impedes traveling on the Way.

Fill your cup too full and it will spill. Sharpen your knife too much
and it will blunt. Acquire too much and you will be unable to protect it.
Climb too high and you will be unable to maintain it. Too much pride
brings on its own disaster. Too much study leads to exhaustion.
More words mean less. Walk too far and you will pass your destination.

The greatest mistake is desiring too much. The greatest sorrow is not
knowing when you have enough. The greatest fault is needing to compete.
The greatest defect is not knowing when to stop.

CHAPTER FIVE

Balance

*Grand Versus Prosaic. How Much Garden? Limiting Factors. Too Much Tilling.
Too Much Watering. Too Much Fertilizer. Too Many Pests. Knowing When to Stop.*

Once upon a time in ancient China there were two men, Shan Pao and
Chang Yi. Chang Yi cared only for the outside. He pursued profit. He
visited every rich or famous person and went to every party and curried
favor and established influence. He bought and sold and schemed and
planned. He ate rich foods and indulged in every physical pleasure.
When Chang Yi turned forty he was wealthy and famous. But his insides
became diseased and he died before he was forty-one.

Shan Pao cared only for the inside. He ate only rice and vegetables and
drank only water. He gave up desire for prestige and position and profit,
and he withdrew to the mountains, where he lived austerely and spent
all his time practicing physical and mental disciplines. At the age of fifty
Shan Pao was still strong and fit and as agile as a youth in his prime, and
his face was as smooth as the face of a newborn baby. But then one day as
Shan Pao strode alone along a mountain path he was attacked and eaten
by a tiger.

Chang Yi cared only for his outsides, and his insides were eaten by
disease. Shan Pao cared only for his insides, and his outsides were eaten by
a tiger.

Grand Versus Prosaic

One season I did some of my garden experimenting on a bit of land that was part of a farm just
acquired by the farmer. When he showed me
around, he proudly displayed his five-hundred-
year plan for the farm. My gardening on that
farm was short-lived. The farmer, unable to meet

operating expenses, lost the farm in less than a
season. I came out one day and found that all my
"permanent" raised beds had been plowed under
and turned into a field of silage corn by the neighboring dairy farmer who had leased the place. So
much for five-hundred-year plans.

In my gardening I often find myself struggling with the conflict between short-term and

long-term goals, and between the mundane and the grandiose. A basic bottom line is that I need my garden to produce all my tomatoes, cooking greens, peas, and summer squash. Beyond that, it's still prosaic to want to produce a good amount of my own staples, such as the kinds of corns that make spectacular cornbread and polenta, as well as dry beans and winter squash. In addition, however, I also want to breed better, more vigorous, more delicious varieties of vegetables, ones that grow and yield spectacularly well under organic growing conditions and under conditions of relatively modest fertility. In my mind's eye, these varieties are exactly the optimal varieties to form the core of Victory Gardens and to help people survive good times and bad throughout the next thousand years. That is about as grandiose as it gets. Ridiculous, really, when you consider that when I started breeding vegetables I had just three tiny beds that added up to no more than about 100 square feet (9.3 square meters) of garden.

In the years when I had little gardening space, I found myself using so much of it for plant breeding projects, trials of different varieties, and other experiments designed to yield primarily information that I often found myself short of vegetables to eat. Some plant breeding projects give you food all the way through, but many don't produce prime food initially. Trials and experiments to evaluate growing conditions and optimal methods produce food, but often much less in the short term than just going with whatever is already tried and true. So there is a basic conflict between the immediate and prosaic—just a decent amount of my favorite fresh vegetables this season—and the long-term and grandiose—better varieties and better food security and resilience for everyone for the next thousand years.

In the short run, my drive to generate new varieties and new information undoubtedly cost me with respect to the simple business of producing my own food. However, the best variety of delicata squash in existence, the one that is most delicious, most vigorous, most productive, and has the biggest fruits and thickest flesh, is 'Candystick Dessert Delicata', and it was bred by myself and my friend Nate France. And the best big *Cucurbita maxima* squash is my own reselected line of 'Sweet Meat': 'Sweet Meat—Oregon Homestead'. And my favorite corn for cornbread and polenta is my own 'Cascade Ruby-Gold Flint'. In addition, selling seed of these and other varieties I've bred now supports all my further plant breeding and food gardening and allows me to lease land and operate at a scale more commensurate with my drive to create new varieties and develop better methods.

At this point, I'm very glad I sacrificed the immediate to the long-term as much as I did. With my own varieties of corn, beans, and squash, and with the eat-all greens garden method I've pioneered and describe in this book, I have much better food that is much easier to produce than was possible before I started breeding plants and experimenting. In fact, gardening would take much less space and be much easier with no further conflicts between prosaic and grandiose . . . if I could just settle down and stop breeding more stuff and wanting answers to additional questions. Which, of course, I can't.

It's hard, though, to get the balance just right between immediate and long-term goals. Until last year I failed to plant fruit and nut trees. My own backyard has shallow soil and is completely shaded by trees on the property of neighbors except for small bits over the septic drainage system. That's a better situation for raised beds of shallowly rooted annuals than for trees. At this point, all my gardening is done on borrowed or leased land, and that doesn't usually permit planting trees. People are often willing to lend you land they aren't using, but not usually to give it up permanently. About five years ago I leased a couple of acres (0.8 hectare) of land near home, however, and still didn't plant trees. Too many other things were happening, and it still wasn't "my" land. And I never quite got around to preparing the ground properly at the right time. There are always plenty of good reasons to put off the longer-term goals. Last year I decided to stop worrying about the "my" land issue. I also decided that if the preparation needed to be ideal, it wasn't going to ever happen. With a few minutes of conversation, I got permission from the owner of the land to plant the trees. I slammed them into the ill-prepared ground and planted my annual greens garden between them. Optimal it wasn't. But the trees lived. It's a start. Life is too short to not immediately plant fruit and nut trees.

How Much Garden?

How much garden do I want? How much *physical gardening labor* do I want to do? That tends to be a very different story. Gardening can be a good place to learn restraint. Beginning gardeners who have enough space to do so usually plant way more garden than they can water or weed. The result is often a catastrophe in which little produce or joy results. Experienced gardeners, no matter how much space they have, generally plant a little more than they can take care of most years. This is reasonable. We gardeners

are the quintessential optimists. We have to be in order to garden at all, considering how many bad things could happen to each seed we plant or plant we tend, and the months of patient tending required before any harvest. So, being optimists, we plant as much as we can take care of in a good year. This means that we have somewhat more garden than we can take care of in an average year, and much more garden than we can take care of in a year in which various other unpredicted things intervene and we have less time for gardening than we had hoped.

During my first several years of gardening I had just a few tiny beds for gardening. There was no way for me to plant too much garden. I just didn't have the space. Restraint was still learned, but it was the restraint associated with setting priorities. Only in later years, when I was already an experienced gardener, have I had enough space to have the humbling experience of planting too much and being overwhelmed by it. I have had my full share of such experiences. Gardening is just so seductive. There are always too many plants I want to grow, too many new things I want to try, too many questions I want to ask (by planting and trying things), and too many new varieties I want to breed. With the right amount of garden, the gardening itself is a joy and labor of love. Too much garden can become a joyless slave driver.

My solution has been largely organizational. The parts of the garden that are nearest the road and irrigation system and easiest to tend are where I focus most of my energy. I plant the less important crops farther from the road and the irrigation system, where they tend to get less care automatically. When I have too little time, I don't spread that too little time over the entire garden and risk tending it all poorly. Instead, I always tend the most important beds and areas first. If at some point the outlying sections become so neglected as to make it obviously the thing to do, I just hoe them down or till them under. I haven't learned how to restrain my gardening so much so that I always have the right amount of garden. In the real world, where we can't predict the other demands on our time and lives, this would probably not be possible. Instead, I have gotten better at organizing so that I can write off parts of the garden with minimal wasted work. That is, I've learned to adjust the size of my garden downward where warranted during the season.

Limiting Factors

One of the things that has most helped me plant the right amount is to think in terms of limiting factors. In my home garden, for example, work in watering (not the actual water itself) is a major limiting factor. My "home" garden is actually a five-minute drive from home, and my watering is with sprinklers and just a hose system. I can run three sprinklers at once. In addition, it's not usually convenient to water and work in the garden at the same time. The watered section would usually need to be one I was working in or walking through, and the soil becomes very mucky when wet. I've found that about half an acre (0.2 hectare) of summer annual crop garden is as much as I can water regularly. This means that any warm-weather annual plantings that need watering need to be limited to that amount of space. The summer greens crops, plus green beans, peas, tomatoes, cucumbers, and squash, all need to go in that space. A small orchard that I

have just planted with traditional generous tree spacing should, beyond the first two years, not need irrigating, so it doesn't have to fit within that half-acre limit. Nor do October-planted overwintering crops such as garlic, fava beans, and overwintering peas, which grow during our winter rainy season. Very early-planted eat-all greens patches can also be in addition to the limit, as they can produce a crop on spring rain.

The size of the squash patch I grow for my experimental and breeding work is limited partly by the need for watering and partly by the amount of weeding it takes. The weeding is easier than for most crops, though, because the plants start off as just a few individuals that are in mostly empty space that is easy to weed initially. Then later in the season after the vines have started running they quickly shade out nearly all the weeds.

I prefer to grow potatoes without irrigation, even when irrigation is available. With potatoes, time and effort spent in weeding and hilling up is the major limiting factor. Hilling up is not only time consuming, but physically very strenuous. Digging up the spuds is also demanding. The size of my potato patch is labor-limited.

The amount of time it takes to harvest a fall-harvested storage crop isn't usually a limiting factor because it's typically a one-shot deal. If weather emergencies threaten, I can usually solicit help from friends for this kind of onetime harvest. In addition, I grow enough different varieties so that my harvest dates are generally somewhat staggered, in part deliberately to ease the amount of harvesting labor involved in any given week.

The time required for harvesting fresh eating crops as well as the timing is a major issue, as is the time required to prepare the crop for eating— the kitchen prep time. I have found, for example, that I won't harvest cherry tomatoes if full-sized tomatoes are available because of the harvesting and kitchen prep work. It takes a lot more harvesting *and* prep time to harvest a salad's worth of cherry tomatoes than full-sized ones. Likewise, I prefer edible-podded peas to shelling peas. Edible-podded peas give much more food for the space and picking labor, and require no laborious shelling.

Timing of the labor also matters. For example, for just my own use, I can keep up with and eat the harvest of a 10-foot (3 m) row of edible-podded peas. However, I won't keep a 20-foot (6.1 m) row picked. Picking pea pods is laborious. I love the peas enough so that I will pick all I can eat fresh. But the harvesting labor involved is too much for me to want to plant extra peas for freezing. So I plant three small patches of peas in succession so I have the right amount of peas all season long instead of too many all at once.

Many people dedicate at least part of their gardens to crops for long-term storage, for root cellaring, freezing, canning, or fermenting. In these cases the time or facilities they have for the processing or storage may be the limiting factor.

Sometimes the need for protecting a crop from pests is the limiting factor. When I gardened in my backyard an important distinction was what I could plant inside versus outside the deer fence. I grew only garlic and very hot mustards outside the deer fence. I could plant more generous amounts of these crops since they did not cost me the more limited fenced-in space. Two dwarf cherry trees planted together so as to cross-pollinate each other and of suitable size to be covered with bird netting may give you more cherries to eat than a dozen full-sized trees unprotected from cherry-loving birds.

Too Much Tilling

Plowing, rototilling, and even just plain digging with a shovel all disrupt soil structure. They also stir and aerate the organic matter in the soil so that it is much more rapidly degraded by fungi and bacteria. If we till without adding organic material to the soil, the overall organic matter content will drop year by year. Soils with low organic matter tend to be hard instead of soft and fluffy and difficult for roots to penetrate. The surface of the soil may crust when it dries out, making it hard for seeds to germinate. The water-holding capacity of the soil drops so the garden becomes more drought-sensitive. In addition, there may not be enough organic material present to provide enough nitrogen for crops, so we may need to add nitrogen-releasing organic material yearly as part of our garden or farming routine. Basically, soils that have a good organic matter content are soft and fluffy and are optimal places for seeds to germinate, roots to penetrate, and plants to grow.

Ideally, it would be nice to be able to grow our gardens without digging or tilling at all. Masanobu Fukuoka, in *One-Straw Revolution* and his other books, advocates exactly that. Most of us have not been able to pull that off. Exactly why Fukuoka could get away without tilling and the many Americans who have tried it can't is as yet a matter for further research and pondering. So far, I have always had to dig or plow or till at least once a year. I generally need to start with plowing or digging to convert old pasture or lawn to a garden in the first place. It's full of vigorous grass or weeds, and just sprinkling seeds on top of unturned earth is a waste of seeds, except when they land on the occasional gopher mound (where the gopher has done the digging). There is simply too much competition from established plants. So I start by turning under existing plants. Then, after growing a year's worth of crops, the ground has usually compacted enough so that the next season's garden seeds don't germinate and grow very well in it without some digging first. In addition, I often have enough of a weed problem so that I need to till to control weeds. I would prefer to dig not at all or only initially to establish a garden or farm field. Instead, so far I have ended up tilling once or twice yearly.

To maintain the organic matter content of the soil in the face of even once-yearly tilling, we generally need to do something more than just turn under the garden crop residues or compost made from them. A garden with its empty non-growing space and its loss of biomass through removal of crops and (in some cases) weeds doesn't produce enough total biomass of organic matter to make up for all that is burned up by even once-per-year tilling. On a small garden scale, we can import and add compost or other organic materials. Grass clippings or leaves from the rest of the yard (or neighborhood) can be used to maintain the organic matter content in the garden. Another garden-scale solution is to lightly mulch most or all of the garden, including the paths. A 2- or 3-inch (5.1 or 7.6 cm) layer of leaves or straw, for example, is enough to protect the ground and inhibit most of the annual weeds, as well as to prevent water loss. (It doesn't stop perennial weeds, however.) By the end of a season of gardening the mulch has mostly been worked into the ground by earthworms, sow bugs, and other critters, and has contributed its organic matter to the soil.

On a farm scale, in order to maintain organic matter in the face of once- or twice-yearly tilling

we must usually import manure or other organic materials and/or use cover crops or green manure crops. (Cover crops and green manure crops are grown to be tilled under instead of harvested so as to replenish the soil's content of organic matter and fertility.)

If you use a rototiller to weed between rows in season, set the tiller to till as shallowly as possible to reduce organic matter loss. Also consider switching to a wheel hoe. In a previous era we used to use a rototiller to till between rows in our home garden. However, when the tiller broke I shifted to a stirrup hoe and a wheel hoe for that purpose. At best, rototillers are loud, unpleasant machines that are hard on the body. Hoeing, when done correctly, can be a relatively easy-on-the-body and enjoyable, even meditative activity.

Rototillers aren't optimal for the initial digging that's needed to convert pasture or lawn to garden. Unless the area is very small, doing the initial tilling with a rototiller is too time consuming and hard on the body. In addition, rototillers simply don't dig very deeply when they are used for primary tillage. Six inches (15.2 cm) or less is standard. Where the ground is hard the tiller may just stir up the surface. Furthermore, you can't necessarily see where the tiller hasn't dug deep enough until long after the tiller is put away. I once started planting squash in a rototilled area that looked fine. And it was, but only the top 2 inches (5.1 cm) of it.

Most gardeners who have a big garden but are tractorless hire the local tractor guy to come plow at the beginning of the season. Then they use the tiller set for shallow tilling just to weed between rows, or to till patches between crops within the season. I hire the tractor guy in spring, but maintain paths thereafter with hand hoes and a wheel hoe. I am trying to get away from fall plowing entirely.

Digging, plowing, or tilling of any sort needs to be done when the soil is at the right moisture level. If the soil is too wet, when it dries you'll have large rock-like clods that don't disperse easily in response to rain, irrigation, or more tilling. Instead you have a cloddy soil all season or even longer. Soil that is too dry is subject to blowing away during tilling. In addition, if the soil is dry and hard, a plow or rototiller won't bite very deep into it. When you turn over a shovelful of soil, if it is at the right level of moisture for digging or tilling it will hold together a little but crumble when hit with the side of the shovel.

Hand digging dry ground is much more difficult than digging optimally moist ground. During our summer dry season, when I want to start a hand-dug garden bed, I water the ground thoroughly, then wait about three days to get a moisture level that makes the digging easiest. When I hand-dig a bed in spring, of course, it's a matter of timing things just right between the spring rains.

Too Much Watering

Too much watering doesn't merely waste water and the electricity involved in delivering it, it also leaches fertility from the soil. This increases your fertilizing costs as well as pollutes groundwater.

Also, if you water so much that the ground is actually soggy, heavy plants like cornstalks may lodge (fall over). All it takes to water too much is setting a sprinkler and forgetting to get back to it.

Ideally, those of us with big gardens would have automatic electrical or mechanical timers on various of our hoses that would turn the water on and off at the correct times without our intervention. I've bought four different kinds, none of which worked. I've also talked with others who have bought more than one kind, none of which worked. I'll keep hoping and trying.

With hand watering the temptation is to water too little. It takes a good bit of water to moisten the ground throughout the root zone of the plants. In my experience, this is best delivered in two or three doses. That is, water the entire area lightly three times. By the time you finish the first round most of the water will have seeped a few inches into the soil, leaving the surface able to accept more water. The total amount of water needed tends to compact the soil or run off if it's delivered all at once.

I know of no way to evaluate the right amount of watering that does not involve digging holes occasionally to see exactly how far down the moisture is penetrating.

With perennial plantings such as trees or lawn, light watering at frequent intervals is less desirable than less frequent but thorough watering. Shallow watering encourages shallow root systems that are less resilient in the face of drought. And shallowly rooted plants are able to scrounge nutrients from a smaller area, hence don't grow as well without extra fertilizing.

Sometimes too much water is not so much about absolute amounts as about the timing. A heavy downpour on cherry trees can cause all the fruit to split, for example. (If you are likely to have heavy rain in cherry-picking season, it's best to plant varieties that are resistant to splitting.) Too much water on many fruits gives you obviously watery fruit with soggy texture and diluted flavor.

Potatoes with too much water late in the season have lower dry matter content and are more likely to become diseased. In addition, the baking varieties might be too wet to make good bakers. (Baking potato varieties have less water and more dry matter content than boiler varieties.) Even if you water potatoes during most of the season, it's best to cut the water during the last month.

If you are growing dry beans, and some pods are dry but most aren't, when the plants are rained on or irrigated the dry pods will absorb water and can then mold; meanwhile, the pods that have started but not finished drying will swell up and the beans split. I especially like the gold heirloom bean 'Gaucho' for dry beans, not just for its earliness, productivity, and flavor, but also for its pattern of dry-down. It goes from having all green pods (and therefore being able to handle watering or rainfall) to completely dry and ready to harvest in about two weeks or less. Most dry bean varieties take longer to dry down, and leave you in limbo much longer.

Pole dry beans usually mature their seed and dry down pod by pod over a very extended season. You need watering or rain to keep the plants growing. But water will cause the dry pods to mold and the beans in the half-dry pods to split. I go through and pick all the dry and half-dry pods before each irrigation. This is most practical with very large pods and beans, such as 'Scarlet Runner', which is not just a beautiful ornamental, by the way. It's one of the most flavorful of all dry beans. It has a meaty red wine flavor, like beef simmered in burgundy.

Too Much Fertilizer

We fertilize for two major reasons. One is to add elements that may be lacking in our natural soils. The other is to replace critical elements as they are depleted by agricultural production. In the Willamette Valley, for example, the basal rocks from which the soils formed are usually very low in phosphorus, often somewhat low in calcium, and sometimes low in sulfur. The organic approach is to add finely ground rock of kinds that contain and slowly release the needed minerals.

A soil test on my leased home garden area, for example, showed very low phosphorus, low calcium, and marginal sulfur. This was no surprise, as the land had been a grass seed farm once, and had been untended for a decade or more subsequently, during which any agricultural amendments solubilized and leached from the soil without replacement. We added the recommended amounts of elements by adding appropriate amounts of rock phosphate, lime, and gypsum, which contain phosphate, calcium, and both calcium and sulfur, respectively. On the farm field where my seed crops are produced, the soil test showed sufficient phosphorus and calcium but a lack of sulfur. So we added just gypsum. What is lacking in a soil reflects both the composition of the basal rock it came from and the agricultural history of the land. Most farm soils in the Willamette Valley have been regularly amended with rock phosphate and lime, but not necessarily with gypsum or anything else to provide sulfur.

Generally, we can add enough of the needed minerals in the form of ground rock to provide for about five years of intensive cropping. We usually can't add more without unbalancing the soil nutrient levels. The first year after adding the ground rock powders, only a portion of the elements are bioavailable. After the first year or two, more is available. By year five or so of agricultural production, the supply of minerals in the added ground rock is low enough so that it is time for another round of mineral amendment. Some soils in some parts of the world are so naturally balanced that they need no initial amendments for optimal agricultural production. However, after some number of years of agricultural production, they will start to need amendments to replace those nutrients lost when serious amounts of biomass are regularly harvested, removing essential minerals from the fields.

Plants can become just as stunted and slow growing and miserable from having too much of various elements as too little. It's much more important to avoid adding too much of something than to not add enough. If you don't add enough, you can just add it next season. Once you've added something, you are stuck with it long-term.

If your garden is large and regular enough to warrant it, do a soil test. Send the soil sample in during the winter if you expect to have the results back in time for spring tilling and planting. Choose a testing service that is oriented toward organic growers, such as that provided by Peaceful Valley Farm Supply. Otherwise all the recommendations will be in terms of soluble chemicals organic gardeners don't use instead of the ground-rock forms of minerals and organic materials they do use. Exactly what you choose to use as amendments needs to reflect the overall situation for your soil as evaluated in the soil test. For example, organic gardeners can supply calcium by adding agricultural lime (ground limestone rock), dolomitic lime (ground rock dolomite), or gypsum (ground rock gypsum). The first two raise the soil

pH; gypsum lowers it. Dolomitic limestone has serious amounts of magnesium in addition to the calcium, so if you need both calcium and magnesium, it's a better way of adding the calcium. If you have enough or already somewhat too much magnesium (as my soil does) you don't want to add dolomite as the source of calcium. Since the pH of my soil is fine the way it is, I wouldn't want to add gypsum without balancing its pH effects by also adding lime. Soil tests can also report on many minor elements beyond the three major ones I discussed whose levels can be too low or high in various soils.

If you are gardening in one or more small beds with individual histories, a soil test on one bed may not apply to the others, and soil tests may not be useful or economically practical. (Tests normally cost about $50 or more. And I know of no home kit that is accurate enough or informative enough to be useful.) Often the best you can do is to simply experiment. Add the proposed fertilizer to a small part of your garden and try it for a season. Let the plants tell you what they think about what your soil needs.

Nitrogen presents a different kind of problem from the minerals I have discussed. It's highly water-soluble. Organic gardeners and farmers usually provide nitrogen for their crops by additions of relatively high-nitrogen-containing materials such as compost, manure, fish fertilizer, or seed meal, or by growing and tilling under nitrogen-fixing legume "green manure" crops. These organic materials release their nitrogen slowly via the action of soil microorganisms. Even so, most of the nitrogen in an organic garden's yearly inch or two (2.5 to 5.1 cm) of compost, for example, has vanished by the end of the growing season. Most gardeners add some nitrogen-containing organic matter every year. Compost, organic chicken manure pellets, fish fertilizer, and seed meals are common choices. (For coverage of how to provide nitrogen without imported fertilizers using everything from weeds to urine, see my previous book, *The Resilient Gardener: Food Production and Self-Reliance in Uncertain Times*.)

You can generally tell if your plants aren't getting enough nitrogen. They both don't grow very fast, and are distinctly and uniformly yellowish. Adding a little nitrogen-releasing fertilizer cheers them and greens them right up within a week or less. Most gardeners quickly learn to add nitrogen-releasing materials to provide enough nitrogen for their crops.

Too much nitrogen is also often a problem, and it tends to be subtler and harder to recognize than too little nitrogen. The amount of nitrogen needed varies from crop to crop and depends as well on the time of year and even the weather. It especially varies with tilling. Tilling encourages the breakdown of organic material in the soil and the release of its nitrogen to the atmosphere. The more you till, the more you have to fertilize. Winter rains or melting snow cover can wash away any free nitrogen so it is not available to spring plantings. In addition, microbial action is slow in cooler weather, so release of more nitrogen from remaining organic material can be slow. So enough nitrogen can be a bigger problem with early-spring organic plantings than with later plantings.

I consider avoiding too much nitrogen such a large issue that I never add the nitrogen-containing fertilizer (the organic chicken manure pellets or the seed meal, for example) to my other soil amendments. I keep them separate and add them separately in amounts that vary with the crop and season. Here are some of the problems associated with too much nitrogen.

Tomato plants with excessive amounts of nitrogen grow very vigorously *vegetatively*. They are slow to flower, however, and slower to set and ripen fruit. The amount of nitrogen that makes most other garden crops very productive can be enough to make tomatoes unproductive. If you have huge healthy-looking tomato plants that are very late in producing or ripening fruit, it's likely to be because they have too much nitrogen. I plan the position of the tomato planting before I add nitrogen-containing fertilizer or till so I can add less nitrogen to the tomato section.

Nitrogen encourages plants to grow fast and produce succulent growth. This may be fine if we are growing summer salad plants. It isn't optimal, though, if we want to overwinter plants. That succulent growth is made up of bigger cells that are more susceptible to freeze damage. And plants full of such big cells are more susceptible to wind damage. When I fertilize crops planted in fall for overwintering I always use much less nitrogen than for even the same crop planted in spring. Spring soils are usually short of free nitrogen because it leaches out in heavy winter rains, and the microbes that release it from organic matter in the soil are not very active. In addition, with overwintered crops I would rather they grow slower but are less succulent and better able to tolerate freezing.

Legumes such as peas and beans form associations with nitrogen-fixing bacteria in the soil. This means that legumes generally don't need as much added nitrogen as the rest of the garden crops. With excess nitrogen the nitrogen-fixing bacteria are inhibited. So you are adding often expensive nitrogen to the land to provide for plants that could supply their own nitrogen if left alone. In addition, by giving the legumes only the nitrogen they need, you are giving most of the competing plants—all the non-leguminous weeds—much less nitrogen than they need. So it's optimal to take advantage of the legumes' unique abilities and not overdose them with nitrogen. Give them less than your other crops, or none at all.

If you fertilize fruit and nut trees with enough nitrogen so that they grow as fast as possible, their wood will be soft and tender and succulent. The trees will be more likely to freeze out in winter. Trees and branches will be more likely to break in response to wind, ice, storms, heavy loads of fruits, or even just their own weight. We often think when we fertilize and things grow bigger and faster that our fertilizing was a good idea. Not necessarily.

Many herbs are adapted to growing in relatively poor soil. They may not produce the best-quality leaves with optimal amounts of their distinctive tasty signature chemicals if pampered and given as much nitrogen as your ordinary garden vegetables.

I also suspect too much nitrogen can adversely affect the flavor and texture of fruits and vegetables, but haven't any solid data on it. I just suggest being aware of the possibility.

Too Many Pests

I love garlic. Successful garlic growing around here requires learning to control gophers. Idealists sometimes counsel a live-and-let-live attitude toward garden pests. "Eschew violence," they say. "The garden is a good place to learn to cooperate with rather than fight nature," they say. "Learn to share," they say. "Just plant extra, enough for the people and the critters too." Lots of luck with that.

Enough garlic for the gophers is *all*. It doesn't matter whether you plant a dozen bulbs or hundreds. Long before it is time to harvest, the gophers will have eaten every single bulb. Cinch traps work much better with gophers than idealism. A friend of mine with several acres of melons to protect has taken to burning sulfur in the holes. It's actually organic. Some dogs and some cats are effective gopher getters, effective enough to control the gophers adequately with no assistance on our part.

Gophers make burrows that can go pretty deep but also burrow near the surface to harvest all your root vegetables. They generally don't make just barely covered runways near the surface that are obvious. Gopher mounds are crescent shaped and are plugged with soil. Moles make obvious barely covered runways, and their mounds are circular and unplugged. Gophers grab root vegetables by the root and munch, eating the entire root as well as part of the stem. They are loud eaters. I can hear a gopher eating my garlic from at least 50 feet (15.2 m) away. A gopher-eaten garlic plant will have just the final bit of the stem and the leaves sticking out of the gopher tunnel. Moles can damage plants by their excavations, but they are carnivores. They are after earthworms and insects, not plant roots. I trap gophers with cinch traps, but I practice tolerance with my moles.

There are lots of other little tunneling rodents that are after insects or earthworms and leave the plants alone. "Mophers," my friend Alan Kapuler calls them—critters that aren't gophers and may or may not be moles, but whatever they are, they aren't bothering our vegetables.

We gardeners generally do plant a bit more of everything to take care of the routine or occasional insect or pest loss. In addition, we also plant many different kinds of crops, realizing that in any given year, all of one crop or another might fail. I always have spotted cucumber beetles, for example. They are the little green beetles with black spots who like cucumbers, beans, squash, and pretty much everything else. They may completely destroy a planting of seedlings if the beetles happen to emerge just when the seedlings are emerging. In such cases, replanting is necessary. Usually, though, the timing isn't so disastrous, and the beetles just chew holes in some of the leaves on established plants without doing enough damage to be a problem. Slugs, birds, and rodents always get some of certain things. Rabbits sometimes wipe out some of my legume plantings at the seedling stage. But just occasionally. So I just replant. Slugs or birds always get some of my tomatoes. But I plant more than enough for all of us, with some to give away besides. Where the assault is casual or occasional, we can restrain our protectiveness and possessiveness and practice casual tolerance.

In some cases casual tolerance is just not workable. It will lose you most or even all the crop, and such losses will not be occasional. They will be guaranteed. In these cases the successful gardener needs to either give up on that crop entirely or be more aggressively protective. In the Willamette Valley with its long, mild winters, gophers and root crops just aren't a sharing situation. Sometimes birds are such a problem with emerging corn that, after you put up the traditional scarecrow and wait a few days to see that it doesn't work, a shotgun and a few dead birds left prominently in the field are the only solu-

tion. (Crows know exactly what a dead crow is, and may desert a field for two or three years if a couple of dead crows are left in it.) Otherwise, the birds get nearly every seedling, and those that are left are too few and far between to produce enough pollen to pollinate each other and set ears. So birds can completely destroy a corn planting. Birds may also attack the tips of the ears when the corn is in the milk stage, opening the ears and causing them to mold.

A cherry or strawberry crop may absolutely require bird netting. Rabbits love pea and bean plants and may eat every one while it is still a seedling or young plant. You may be able to tolerate rabbits or have to trap or shoot them depending on the amount and significance of the losses. I never use killing traps in open areas where they may harm a neighbor's cat or, for that matter, feral cats, all of whom are welcome in my gardens. In open areas, if I must trap I use live traps so I can release any accidentally caught cats unharmed. I shoot live-trapped pests in the trap; some people dump the trap in water and drown the pest. You should kill the live-trapped pest animal, not release it on someone else's land. If you think you are doing the world a favor by releasing problem animals, try asking the property owner's permission first to see just how much he or she wants your problem.

When I was gardening in my backyard, I had a huge problem with slugs. I got ducks. They love slugs. There is a major section on using ducks to control yard and garden pests in my book *The Resilient Gardener*.

Large critters can be a large problem. A deer or two in a small garden can destroy a lot overnight and the whole garden in a week or two. A herd of deer wandering through can eat pretty much everything in a single night, the squash and tomato plants along with the fruit. Anything they haven't eaten initially is pretty much just because they aren't familiar with it and need to taste it a little first, or they simply haven't gotten around to it yet. They'll eat it tomorrow. In my experience, hungry suburban deer sooner or later will eat everything you grow that can be eaten without cooking except horseradish, garlic, and hot mustard greens. Scaring devices have only temporary effects. Repelling odorants wash off with rain or irrigation. The approaches that work long-term are deer fencing, a dog kept loose outdoors inside a fence with access to the garden, and venison stew.

Wherever gardeners meet we discuss our common enemies, the pests. We bond over our tales of trials and tribulations. One of my friends has a hard time growing a garden because he lives near a river and there are too many nutria. Nutria are rodents. Think in terms of a huge aggressive rat weighing up to 30 pounds (13.6 kg) that likes just about all garden crops, including even the stalks of corn. Some Australian gardeners these days have troubles with urban kangaroos. Another gardening friend of mine reports that bears love pot. At least, the mother bear and cub who ate his mature pot plants and went to sleep on the remains while he watched them and wept did. Apparently stoned bears don't shoo. I'd be delighted if Sasquatches exist, but I'm pretty sure they don't. If Sasquatches did exist, I'm sure one of my garden friends would have reported on how bad it is when Sasquatches get in your garden.

Knowing When to Stop

The garden is one of the best places to learn about knowing when to stop. There are two kinds of knowing when to stop—immediate and ultimate. Immediate Knowing When to Stop is a matter of being tuned into and respecting your body, your land, your plants, all the rest of life, and the weather. Immediate Knowing When to Stop tells you that when you are hoeing weeds and it starts raining and the ground gets too wet to hoe, it's time to stop. And when you have labored enough so that further labor is likely to produce pain or soreness or loss of productivity the following day, it's time to stop. It is definitely time to stop if you feel your heart struggling or pounding. My dad, while working in the yard, tried to complete one task too many. He had a heart attack. He never completed that or any other task. If it's so hot that you are in danger of heat exhaustion or heatstroke, it's definitely time to stop. Prudence trumps completion when it comes to your health and safety. Most of the time, though, it's just efficiency or practicality or inconvenience at risk, not our lives. One thing I seem to have to learn repeatedly is that when I am gardening and it has gotten so dark I can't see the plants, it really is time to stop.

Ultimate Knowing When to Stop is more complex. Ultimate Knowing When to Stop encompasses all the experience and wisdom you have with respect to gardening as well as the entire rest of life. It involves planning, predicting, practicality, and organization. The garden encourages us to learn both the Immediate and the Ultimate. In the garden, as in all life, you have to learn both in order to fully practice either.

One spring, for example, I find myself needing to hand-mix a combination of lime, rock phosphate, and gypsum to be spread on a new quarter-acre (0.1-hectare) bit of borrowed garden land. I dump the minerals on the driveway bag by bag, move them around with a broadhoe to mix them, then shovel them into tubs with amounts I can carry in the field. The moving around isn't difficult, but it's moving lots of pounds of material. The shoveling into tubs is hard on me. It's hard on the back and knees. I can hoe for hours, but shoveling is just not my thing.

If I dump all 500 pounds (227 kg) of ingredients on the driveway on a day when it is due to rain later, after I have pushed the minerals around enough to thoroughly mix them my body will start to complain really seriously. I know. I've done it. More than once. In the process, Immediate Knowing When to Stop soon tells me that I am risking back injury to continue. Immediate Knowing also tells me that if I continue my arms are going to be so tired and sore that I probably won't be able or willing to spread much of the mix later today or tomorrow anyway, so the whole point in being in such a rush over the mixing would be defeated. However, if the dry minerals get rained on, they become lumpy, difficult to mix, and difficult to spread. Immediate Knowing When to Stop might tell me to stop, but I can't, really. Not without consequences. Immediate Knowing When to Stop only carries us so far.

Ultimate Knowing When to Stop does not go and dump 500 pounds of mineral ingredients on the driveway all at once in the first place. Ultimate Knowing When to Stop takes into account the needs of the body, mind, and spirit. Ultimate Knowing does not create emergencies. It does not dump *any* minerals on the driveway without taking into account and honoring the weather. Ultimate Knowing also considers joy. It realizes that mixing 500 pounds of minerals is really just not much fun, even if rain isn't threatening. With 500 pounds to mix, I look at the mostly unmixed pile and cringe. Most of the work is grim because of how much of the work is left. I need to take breaks, and am tempted to take more and longer breaks than I need.

These days if I need 500 pounds of mineral mix, I mix it in four small batches of 125 pounds (56.7 kg) each on different days. Each batch is so small I feel almost done right from the beginning. I have no desire to take any breaks. I feel just as much pleasure of accomplishment as when I mix a bigger pile, and considerably more pleasure by the time I've finished the 500 pounds. And after I'm done with each small batch, I've just had a mild, pleasant workout spread nicely over arms, back, and legs, with nothing being tired or sore. It's the right amount of work for me physically and emotionally. Ultimate Knowing When to Stop organizes the work so I most enjoy it, and so that stopping at the right time comes naturally.

*In seeking knowledge, day by day something is added. In following Tao,
day by day something is dropped. Day by day you do less and less deliberately.
Day by day you don't do more and more. You do less and less and don't do
more and more, until everything happens spontaneously. Then you act
without acting, and do without doing, and achieve without forcing.
And nothing is done. And nothing is left undone.*

*Muddy water, when still, gradually becomes clear. Be still.
Let your mud settle and your mind clear.
Wait quietly until the right action comes naturally.*

道

CHAPTER SIX

Non-Doing

Daring to Not Do. On Not Tilling, Digging, Mowing, or Tending Absolutely
Everything. Twenty-Four Good Places Not to Plant a Tree. Seven Reasons
Not to Chop Down a Tree. Thirty-Seven Reasons for Not Planting
Various Vegetables. On Not Planting Purple Flowers in Front of an
Orange Brick House. Flower-Patterned Shirts Attract Bees.
A Weed by Any Other Name Is Usually Still a Weed.

Daring to Not Do

There are three reasons to do something: It is the right thing to do, it is the right time to do it, and you are the right person to do it. Usually, it isn't, it isn't, or you aren't. Gardening books and magazines usually focus on doing. They report the positive—things that worked at least once for someone somewhere on the planet. That is only part of the story. We gardeners are an inventive lot. We are capable of thinking of lots of other things to try that we have never seen anybody do or write about. Many of these other things have undoubtedly been tried repeatedly by gardeners in many times and places, and have failed to work for every single person who tried them. For everything that at least sometimes works, there are many-fold other things that never work. I have discovered quite a lot of these. You probably have too.

It is easy to fool ourselves into doing unnecessary things. Our brains are more designed to see overall patterns and stories than to accurately see individual elements. We see a wavy line shape and our brains scream "Dangerous snake in the path ready to strike at me!" We freeze or jump aside long before our eye and brain have time to more accurately report on the individual elements and build them into a more accurate story—there is just a dead stick in the path. Our mental bias toward leaping to perceiving patterns is undoubtedly a survival trait. We can afford to startle erroneously over hundreds of dead sticks if our pattern-seeing bias allows us to respond as fast as possible to the occasional real poisonous snake.

As part of our bias toward seeing stories, we tend to interpret sequences of events in terms of causation. We see cause where there is only correlation. I think we also have a hardwired bias toward intervention. So we usually try to solve problems by doing something, even where doing nothing might be the best option. We often don't give a problem any chance to resolve by itself without our intervention (and added work). Then when we have done something, we interpret any changes in the situation as being caused by our action. We may then add that action to our list of gardening chores, and do it from then on without ever testing whether it is useful, let alone necessary. Many traditional gardening tasks are unnecessary interventions that are both laborious and counterproductive.

When we gardeners intervene or try something, ideally we should use controls. We should try the new thing on only part of the crop. We should dare to not do on part of the planting. This requires objectivity and humility. All too often it is enough easier to do something on every part of the crop and forget about controls. That may be easier in the short run, but harder in the long run, when our built-in biases end up adding unnecessary work to our gardening for the rest of our lives. The agricultural patterns we inherit are mixtures of components, some of which are necessary and optimal, and most of which are either unnecessary or suboptimal. This is part of why gardening and farming are so much fun. Given just one concept—the idea of not doing things on part of the planting whenever we try something new—we can easily make new advances.

Energetic gardeners are easily seduced into doing unnecessary things. We somewhat lazier gardeners are often saved by our laziness. When it is time to do the work we find ourselves instead lying in the grass in the shade drinking lemonade and trying to think up reasons why the work may be unnecessary. There is nothing like a lazy gardener when it comes to daring to not do. Sooner or later, we are likely to try not doing just about everything. In addition, we try not doing quite a lot inadvertently. We *meant* to do the task . . . really, we did. But when the time came, we just didn't get around to it. Sometimes, delightfully enough, things work out just as well or better when we skip that task. That is our most magnificent reward. Nothing is so happy as a lazy gardener who has discovered a bit of gardening work that doesn't matter.

The classical Taoist term *wu wei*, usually translated as "non-acting" or "non-doing," incorporates what tend to be two different concepts in the West. The first is absolute non-doing, just plain not doing things. The second aspect of *wu wei* is closer to "efficiency"—maximally competent doing—doing that gives the maximum effect for the minimum effort because every unnecessary action has been eliminated, and the little doing that is left is done optimally. I cover efficiency throughout this book. In this chapter, I address just plain non-doing. (I suggest other gardening things to just plain not do in the section on Selective Sloppiness in the Labor and Exercise chapter of *The Resilient Gardener*.)

On Not Tilling, Digging, Mowing, or Tending Absolutely Everything

All my favorite places to walk and play and be when I was a kid were the places outdoors that no one was tilling, digging, mowing, or tending. They were the vacant lots, the swamps too low and wet for houses, the strips of woods along railroads, the abandoned edge land between housing developments and farms. In most cases, every square foot of land some adult property owner actually starts actively tending is one less square foot of land for neighborhood children to play in and thrive on. It's also usually one less square foot where anyone can most joyously go for a walk.

Wildlife, including insect pollinators, often need these unmowed, untilled, untended, or minimally tended places. Ground-nesting birds need unmowed pastures or meadows to hide their nests. Such places help preserve the ecological health of the tended crops and land as well as the biodiversity of the planet.

Once you till or dig, you usually have to continue tilling or digging. Disturbing the ground creates space that begs to be filled by disturbed-ground-specializing weeds. So you then have to continue tilling or digging in order to prevent takeover by the weeds, weeds that are often less desirable than whatever you tilled or dug to eliminate. Once weeds take over and fill the soil seed bank full of weed seeds, many styles of gardening can become less workable or even impossible. You may be able to grow vegetables only by transplanting large transplants and then mulching heavily to suppress weeds, for example. You may not be able to direct-seed most things. And weeding can become something that has to be done repeatedly during the season, whereas with less weed-seedy soil one or two weedings might be all you need for the season, and even those might take minimal effort.

Lawn or pasture can be maintained by mowing, which is generally easier than digging or tilling. Pristine lawn, of course, takes careful tending. But just basic grass in a crude lawn or pasture in all but the driest regions can be maintained with no fertilizing and watering, and with just a couple of mowings per year. Leave the lawn or pasture or grassy meadow alone until you are ready not just to till it once, but to till it and then tend it more intensively from then on.

If you have a tiny backyard, it's reasonable to turn it completely into garden and orchard and some lawn, even pristine lawn, if that is what you need. If you have the luxury of owning or controlling more land than you need for these things, you can afford to exercise restraint. You need not intensively manage every square inch of the land just because you can. You can leave room for wildlife, windbreaks, woodlots, wetlands, places where it's great to go for a walk with a dog off leash, and even room for human children. The land got along for four and a half billion years without

your supervision. Some of it can probably survive for a few years more.

Twenty-Four Good Places Not to Plant a Tree

Unplanting a tree is much harder than planting one, so of all gardening things it's good to learn not to do, at the very top of the list is not to plant trees in the places where they shouldn't be planted. Every property has many good places to not plant trees or bushes. Planting any tree or bush requires digging a hole. A good place not to dig a hole for any reason is on top of underground power or gas or water lines. Check with relevant utility companies before digging.

One of my favorite places not to plant a tree or bush is over the septic drainage field. Most trees and bushes have roots capable of expanding so as to clog the drainage field. Then you get sewage overflowing into your yard or even backing up into your house and coming up and spurting out of tubs and toilets. The tree or bush must then be dug up, the drainage field rebuilt, and the house decontaminated. The upper parts of septic drainage fields are often obviously marked by the greener, taller grass growing over them. The complete network should appear on the property title or legal description. Note that adjacent houses may have septic drainage fields and easements for them that extend onto your land. These should also appear on the land title.

Grass has roots that don't expand and clog septic drainage fields. Here in the maritime Northwest, septic drainage fields in heavier soils work not by drainage so much as by the grass over them removing water and nutrients as it grows. Grass is an essential part of the functioning of the system. Bamboo is a grass. I think this means you could have a bamboo grove over your septic drainage field. But since I haven't seen anybody actually do this, I make no promises.

Another great place not to plant a tree or bush is anywhere where there isn't going to be enough room for it when it is full-grown. The bush planted right next to a path is going to block the path given time, or at least try to. You will find yourself fighting with the bush every year over who gets to use the path. This makes for much pruning that you would not have needed to do if you had positioned the bush more appropriately. The same applies to planting trees too close to driveways, sidewalks, or streets. Or under power lines. Don't plant trees and bushes where you will be unhappy if they don't grow and pissed off at them if they do.

Falling fruit or nuts can dent cars. I once lived in a house that had two giant walnut trees near the street, and I made the mistake of parking under them one nice fall day. I had no idea how hard those nuts could hit when dropped from a tall tree. I ended up with dozens of little round dents across the entire roof of my vehicle. So near a street might not be the best place to plant fruit and nut trees, particularly where people are dependent on on-street parking. Close to your greenhouse where they will be able to bombard it when they get to be full-sized isn't a good spot either.

If the main point of planting fruit or nut trees is to get the fruit or nuts, don't plant such trees where they will be impossible to harvest. Land can be too steep to harvest comfortably. If your tree's branches extend over your neighbor's fence or property line, it's problematic as to who should get the fruit, and you may not have the access required to tend the overhanging branches prop-

erly or to harvest. Fruit or nuts that land on the roof of your house will usually be wasted because it is too inconvenient to harvest them; and there will be debris and rotting fruit that you will need to clean off the roof each season.

Some trees when planted near sidewalks or streets buckle and break the pavement with their expanding roots. Willows are particularly notorious for this.

Trees or tall bushes planted near a public street, especially close to an intersection, can block a driver's view and thus create a safety hazard. Your city may have ordinances specifying how high shrubbery planted in such places can be. Whether you have such laws or not, plant only low-growing things in areas where public safety requires visibility.

Many people plant only low-growing bushes or no bushes near their home because they make the home a more attractive target for criminals. Tall bushes near your windows create a cozy spot for robbers or home invaders where they can hide out of sight while breaking in. Tall bushes near your door mean that you can be ambushed when entering or leaving your home.

Trees or tall bushes planted near your home can provide a route for critters such as raccoons to get on your roof, where they may pull up shingles and make a hole in your roof that lets the rain in and causes a big leak over your bedroom, which in turn causes your bedroom ceiling to cave in and dump hundreds of pounds of ceiling and soggy insulation onto your bed. It's the roofer who tells you that you need to chop down the bushes or the coons will quickly tear up the temporary tarp as well as continue to threaten the new roof once it is installed. This experienced roofer also says that the main reason animals get into attics isn't a lack of grates that protect vents, but instead is trees and bushes near houses that make it easy for animals to get on the roof in the first place. If the critters can get on the roof, they'll get in the attic somehow or other.

In areas where wildfires are a major hazard, not planting any bushes or trees anywhere near your home is a major method of minimizing the fire hazard to your home.

In urban and suburban areas people often tend to think that any fruit or nuts on trees planted in the strip between the sidewalk and the street are free for the picking. You own the land and pay taxes on it, and may have planted the trees, and may be tending them carefully. Nevertheless, much of the public tends to consider this area basically public property to some extent. This attitude was actually legally sanctioned in Jewish law in biblical times. People were allowed to harvest grain or fruit adjacent to and within a specified distance of any public right-of-way. This wasn't considered theft. It was part of the repertoire of social mechanisms for helping to ensure that the poor and landless did not starve. If you have no excess land and will need all the food you grow yourself, plant fruit and nut trees in your yard at least a little distance from the public right-of-way. (Ideal is in the backyard. Second choice is a fenced part of the front yard.)

If you have enough space, consider planting fruit trees near your public right of way, and put up a sign indicating that you want to share. This will enhance neighborhood esprit as well as food resilience. I once lived near a house with a beautiful flower garden near the sidewalk that sported a sign saying FLOWERS. HAVE SOME. Bouquets of flowers in my house isn't part of my pattern, but that sign cheered me up every time I walked by it.

A much-traveled sidewalk in front of your house is a good place to contribute any excess garden vegetables. Nothing cheers a sidewalk up like a table with several boxes of vegetables and a sign saying I HAVE PLENTY, HELP YOURSELF. I'm in favor of giving away excess produce rather than selling it unless you really need the money. Modern life has become overly monetized, I feel. In many places in other eras a person could travel anywhere and ask for hospitality—food and shelter for the night—from any household. This meant that people could exist and even travel without needing to have money or be affluent. The more we can do to help create patterns of living that are less dependent on money, the more freedom and security we will have.

Most fruit and nut trees thrive best in full sun. Many vegetables prefer full sun, so the sunniest spot is the best spot for the vegetable garden. On many properties, sunny spots can be at a premium. Don't plant an inedible tree or other ornamental in the one spot that has enough sun for fruit trees or a vegetable garden. If you are short of sun, don't plant shade-tolerant plants anywhere but in the shade.

The tree you are thinking about planting in that spot—what is it going to shade out when it is big? The only spot that is sunny enough for fruit trees or a vegetable garden? If so, that's a good place not to plant that tree.

Vegetables don't compete well with tree roots, and tree roots extend out to about one and a half times the extent of the branches. (This is a more recent figure than the one time-radius that's usually mentioned.) So, ideally, you really don't want trees near your vegetable garden, even if they won't shade your vegetables. In a small yard, however, toward the outer edge of the tree's root zone might be exactly where you will want and need to plant your bushes and vegetable garden. On the smaller property, you don't usually have the option of going for the ideal. You settle for the workable.

It's obvious that bushes planted right in front of the water spigot will block access to the spigot. Some other access issues aren't so obvious. There are fireplaces in my house, and I would love to use the chimney to install a woodstove. There is free wood to be had from neighbors when large trees fall down. There is even a spot where I could put a woodshed or woodpile. However, I can't reach that spot or anywhere on most of the property with a truck because the former owners built a concrete retaining wall to hold up plantings of inedible bushes across the only possible access route. So there is also no way to get amendments like sand, compost, or leaves down to the main part of the property. It's easier to garden if you can reach the garden with a pickup truck to bring in sand, compost, or leaves. It's easier to maintain a property when you have full access to it. Outbuildings such as barns and sheds are valuable. If you don't have one now, you very well might want one in the future. Don't plant trees or bushes in such a way that there is no room for or access to an outbuilding you will need later.

If the tree or bush is going to need irrigation, don't plant it in an area that will be difficult or impossible to irrigate.

If the tree or bush is going to need irrigation, don't plant it with other plants that don't need irrigation, or that need much more irrigation. Your labor and water will be best conserved by grouping together plants with similar water needs.

If you sprinkle a dozen fruit trees in a random arrangement over your entire yard, you will have a yard that is difficult to mow. It's usually easier to maintain your property when your fruit trees are in an orchard or at least mostly in groups of some kind.

Don't plant the pollinating variety far away from the tree it is supposed to pollinate. The two cherry trees of different varieties that need each other as pollinators should be planted adjacent to each other. Some plum tree varieties need another variety as pollinator; even the varieties that are self-fruitful usually set fruit better if there is a different compatible pollinator variety nearby. Almost all apple varieties require a pollinator of a different variety. Unless you have a combo tree with multiple varieties grafted onto it, you need a pair of apple trees, not just one.

If your neighborhood is overrun with hungry deer and you are planting a deer-delectable tree, outside the deer fence is a good place not to plant it.

Every place where you plant an inedible tree or bush is somewhere where you won't be able to plant an edible one. I suggest not planting anything inedible anywhere at all until you have planted or reserved the space for planting your vegetable garden, fruit and nut trees and bushes, and all other edibles. Most "ornamentals" are good yard plants partly because they are inedible to everyone, including most wildlife. A yard full of ornamentals makes for a relatively sterile ecology. Fruit- and nut-bearing trees and bushes provide food for wildlife as well as people and enhance the food resilience of your family and community. On the smaller property it is reasonable to avoid ornamentals entirely and plant only trees and bushes that will produce food.

Many ornamental trees and bushes produce large amounts of highly allergenic pollen. Male cultivars of many dioecious trees—those that have male and female flowers on separate trees—are notorious for it. Wind-pollinated plants depend on producing huge amounts of pollen, and are often associated with allergies. Plants pollinated by bees or insects usually don't produce as much pollen; in addition, their pollen is usually sticky and is less likely to become airborne in large amounts. These are just generalizations, however. In addition, pollen is not the only plant-associated cause of allergies. Just as some people are allergic or reactive to perfumes or other volatile chemicals, some people are allergic to the odors of particular plants. If you or anyone in your family has problems with asthma, hay fever, or other related allergies, I suggest buying a copy of Thomas Leo Ogren's book *Allergy-Free Gardening* and consulting it before planting any tree or bush. This book rates hundreds of garden and ornamental plants right down to the individual cultivars on a ten-point scale as to their potential for creating or exacerbating allergies. Nowhere at all is a good place to plant a tree that is going to make you sick.

Seven Reasons Not to Chop Down a Tree

It takes years to grow a mature tree. When you first move into a new house, you are unlikely to understand all the functions of each of your mature trees. Wisdom suggests not destroying what we don't understand. It's best not to chop down any mature tree until you have lived with it for at least two or three years and have had a chance to notice its virtues.

Appreciate and preserve the deciduous tree growing south or southwest of your house that casts shade on and helps cool your house in summer, then drops its leaves in winter and lets the sun and light in.

Trees and bushes and woods can act as windbreaks. If you cut them down, you can change the pattern or intensity of wind on your land.

Small patches of woods can provide fence poles, bean poles, firewood, and emergency cooking fuel. And they are great places for wildlife to thrive and children to play.

Trees, shrubs, bushes, brambles, and woods can create visual or physical privacy. When homes were built on the property below mine, I found I had to fence across the lower property line to keep kids and dogs from tromping through my garden and upsetting or chasing my ducks. However, I didn't have to fence the side with the wild blackberry thicket. It provided all the blackberries I could eat in August as well as physical and visual privacy. That thicket completely blocked all view of the neighboring house. The thicket was, alas, actually on the neighbor's property. When he chopped it down our houses were totally visible and naked to each other.

Think several times before chopping down productive fruit or nut trees unless they are simply too diseased. If you don't like the varieties, you can graft the trees over to varieties you do like. If you have more fruit than you need and find it a nuisance to harvest or clean up after, there are probably others who would love to have that fruit, and who would be happy to gather it. Contact a local gleaners' group, or just spread word around in your neighborhood. One place I lived in Corvallis was a few houses away from where the landlord lived with two beautiful Italian prune-plum trees he didn't harvest. He let me harvest all the fruit. I went over every three days and shook the trees and gathered all the fruit. The fruit fell just short of ripe, hard enough to be undamaged by the grass, just perfect for ripening in monolayers in flat cardboard boxes indoors. I ate those plums as my main carbohydrate in plum season and blanched and froze the rest for winter. Fruit and nut trees, even if the fruit is actually wasted most years, contribute to the food resilience of a community.

Honor the tree that can comfortably support a swing. Preserve the tree that is just right for a tree house. Cherish any pair of trees that can support a hammock.

Thirty-Seven Reasons for Not Planting Various Vegetables

There are hundreds of vegetable crops and thousands of varieties I have never tried. I tend to want to try them all. If I plant too many new things, though, I fail to evaluate them well enough for the planting to be a fair trial. When it comes to my regular tried-and-true crops and varieties, I always want to make sure I have enough, so I tend to plant too much. But there is only so much land and so much time. The more I get carried away in planting the less good a job I do on everything. I know this. I tell it to other people. I even write about it in books. However, over and over, each planting season I plant things I've already figured out many times before that I don't need to plant, and plant too much of some of the rest. These days, I keep two kinds of lists to help me exercise more restraint in planting. One is the ordinary list of all the new things I want to try. The other is a list of reasons not to plant things or to plant less of them. Before planning or planting anything (or ordering seeds), I review *both* lists.

Having both lists helps me best set priorities and create a garden that gives me the most of what I want for the space and work involved. My list reflects my own personal preferences and idiosyncrasies, but you'll undoubtedly be able to adapt it to reflect your own. So here is Carol Deppe's Secret Second List—Thirty-Seven Reasons for Not Planting or Growing Less of Various Vegetables.

1. I don't like the taste. It doesn't matter whether you call them beet leaves or chard, or claim the greens of this particular variety of them are especially mild or sweet. They never are. And I don't *care* how well it overwinters. Part of the reason it overwinters so well is nothing else likes to eat it either.

2. The kids (husband, wife, others in the household, ducks, chickens, rabbits, cow) won't eat it. My laying ducks love kale or leafy radish greens, *and* these are easy to harvest for the flock. The ducks won't eat the blazing-hot 'Green Wave' mustard greens. So I cheerfully err on the side of planting too much kale or leaf radishes, but restrain myself and err on the side of underplanting when planting hot mustards.

3. It doesn't grow here.

4. It's the wrong variety. It's a long-season variety, and I need a short-season one. It's for fall planting, and now is spring. The variety is bred for the radish root and makes few and small leaves; I want the leaves. It is a popular variety nationally, but needs more heat or less cold than we have in summer. It needs a longer season than we have. It is a popular variety nationally, but doesn't do well here because it isn't

resistant to the appropriate diseases. The packet is in a rack in a big-box store where the varieties are selected on a national basis with no regard for what actually grows and performs best here. And the seed quality on such racks is often what my friend Steve Solomon calls "sweepings-from-the-seed-room-floor grade."

5. It's the wrong time to plant it. If I wanted that, I should have planted it earlier.

6. It needs to be in full sun. The spot I'm planting isn't.

7. It won't do well on the watering regimen I'm going to be using in that spot (which is appropriate for the other plants in the area).

8. It's too tall for that spot. It will shade out other plantings, or block the sprinkler patterns so other things don't get enough water. (A trellis of pole beans blocks a sprinkler water pattern almost completely, creating a dry strip behind it.)

9. It's too big. If I plant it I won't have room for many other things I care about more.

10. It's too little/low/droopy. The leaves would get all muddy. And muddy salads and cooked greens taste awful. I have already ascertained that I'm unwilling to wash my leafy greens. I need my greens to be big enough and grow upright enough so that they stay clean enough so that I can eat them without washing.

11. It would need to be started indoors under lights. I don't like starting things indoors under lights. I'll only do it for things I like a whole lot and where it is the only option. In my garden, unless you are a tomato you had better know how to germinate and grow from direct seeding.

12. It grows too slowly. Before it's big enough to harvest I'll have stepped on it, forgotten what it is, or changed my mind about wanting it.

13. It grows too fast and spreads. If it ever got away, I'd have nothing else. And it's notorious for getting away. If I want that, I'd better wait until I can put it in a bed by itself. (Horseradish, sunroots, comfrey.)

14. I can buy ones just as good (or better), and they are affordable, and it is hard for me to grow. And it isn't all that nutritious. (I'd rather buy the occasional celery I eat than try to grow it.)

15. Picking, shelling, cleaning, or preparing it is too much work. It doesn't matter how productive it is if getting it from the plant to the plate is too much work. I've found I won't pick the cherry tomatoes if full-sized tomatoes are available. Likewise, I tend to not pick shelling peas because I resist the amount of work involved in the shelling. Edible-podded peas give much more food for the same amount of space and for lots less work.

16. It yields too poorly for the space it takes.

17. It doesn't make a good gift. Most non-gardeners are much more enthusiastic about a gift of tomatoes than zucchinis, and don't even know what to do with mustard greens. I most happily err on the side of planting too much when I know the excess can be joyfully shared. Part of the fun of gardening is having prime gifts to give. Prime edible-podded peas are received joyously, but are laborious to pick. Prime fruit is usually received happily. But not blemished fruit. Most people these days aren't used to

blemished fruit and don't know how to deal with it. It's hard to beat full-sized tomatoes for gifts. They are received joyously and are easy to harvest in the larger amounts needed for gifting.

18. It's too fussy. If I looked at it crossly it might die.

19. Nobody around here grows it. They might know something I don't.

20. Everyone around here grows it. I could swap something else for it, or stand outside somebody else's garden admiring theirs till they give me some.

21. Does it meet the WIRHaMoTo standard? "Would I rather have more tomatoes?"

22. It tastes very similar to something else that grows better or requires less work. Sweet potatoes taste similar to winter squash. But winter squash are much easier to grow and harvest. You can plant from seed instead of needing to create and transplant slips. You can lift the produce right off the top of the ground where it presents itself cheerfully instead of having to dig it up laboriously. Winter squash also keep well under my storage conditions, constituting a carbo-hydrate staple I can eat fall to late spring. Sweet potatoes don't store well here with-out an elaborate temperature-controlled curing process.

23. People will steal it. I won't get a one. (Strawberries adjacent to a much-traveled sidewalk.)

24. There are too many problems with produc-ing it organically, which is how I garden. It's a full-season radish root crop variety, and under my growing conditions such roots get riddled with little moldy worm or insect

On Not Planting Purple Flowers in Front of an Orange Brick House

Many ornamental trees and shrubs are ornamental by virtue of having orange, red, yellow, or purple foliage at certain times of year. These will clash with an orange- or red-brick house. So will a bed of orange, red, yellow, or purple flowers. White flowers can seem boring when considered in isolation, but they can look glorious against a brick house or wall. Purple and orange flowers might not look as nice next to each other as either looks next to white flowers. When planting, consider the background and neighboring plants and how they would go with the prospective new plant in each season.

tracks, for example. I don't relish chopping out the small bits of good radish between all the lines of mold. So if I want clean radishes I grow the fast-developing kinds that produce an edible root in forty days or less, not the full-season daikons.

25. It's a perennial vegetable or herb that should go in a permanent bed, not in an annually tilled part of the garden.

26. It's especially great for processing or canning, which requires hours standing over a hot stove in August when it is way too hot for that, and when I don't have the time anyway.

27. It will produce its entire yield in August just when I'm too busy to even harvest it.

28. It will produce its entire yield in August when I have plenty of other things I like better.

29. It's too fuzzy. If I'm supposed to eat it raw in salads, fuzzy stuff need not apply, no matter how many other people say they like it in salads.

30. It's too thorny.

31. It would need some stakes or poles or something. I probably wouldn't get around to putting them up. And it doesn't do well sprawled all over the ground, which I have already ascertained several times before. I'm willing to do the work of staking peas, pole beans, and tomatoes. Most everything else had better be varieties that know how to grow without support.

32. It's poisonous. (This means "poisonous to me," even if it isn't poisonous to others. I can't eat wheat, for example. Some people can't eat tomatoes or potatoes.)

33. It's "dual purpose." Sometimes this means it isn't *very* good for anything.

34. The seed catalog says it's "very tasty in pies." That often means it's too acidic, astringent, bitter, or tasteless to eat fresh, but if I really want to grow it, it *is* edible— if I put enough sugar on it and bake it in tasty dough—in which case it will taste like baked sugar-dough, which isn't bad at all, though perhaps it would be better by itself.

35. Slugs (bugs, birds, gophers, deer, elk) love it. They would eat every one.

36. It isn't resistant to powdery mildew (enation mosaic virus, plagues of locusts, flying saucers), which is/are common in my area (or might occur, you never can tell).

37. Giant Sasquatches might be attracted to it and come steal it and leave nasty footprints all over the garden.

Flower-Patterned Shirts Attract Bees

In an earlier era of my life a flower-patterned shirt was my favorite shirt to wear when gardening. It was amazing how hard the bees worked to figure out where I kept my pollen. Down the neck maybe? Inside the front flaps where the two sides of the shirt come together? Up the sleeve? It wasn't a matter of a buzz-by or a brief landing and inspection either. It was often a prolonged search with attempts to enter every possible cranny. When a bee does successfully find and enter a cranny, the shirt is likely to rub against her, and she is likely to sting. Clothes that are bright yellow seem to be very attractive to bees; orange and red also appeal. These days I prefer solid green or blue clothes for gardening, and I avoid the patterns and colors most often used to signal an interspecific promise of pollen or nectar. If you dress like a flower, don't complain if you get treated like a flower.

A Weed by Any Other Name Is Usually Still a Weed

It's usually a good idea to avoid planting invasive weeds. People in the American Southeast could have gotten along just fine if no one had ever planted kudzu. Some beautiful or useful plants can also be capable of escaping your control and taking over your property and that of your neighbors. Running bamboos tend to run, for example. Plant one without proper control measures and sooner or later your neighbors will be unhappy with you. Even people who love comfrey admit that if you plant comfrey somewhere, it is going to be there forever.

"Edible weeds" are still weeds. We call them "edible weeds" rather than crops because they are not as controllable as our garden plants and, in addition, produce considerably less or less desirable food for the space than our ordinary domesticated garden crops. I appreciate and eat serious numbers of pounds of garden weeds each year. I have been known to plant dandelions and to hoe around lambsquarters and purslane early in the season. However, all of these weeds produce limited amounts of food compared with the same space devoted to most other garden crops. Their big advantage is in already being there without your having planted them. But planting or weeding around them is fraught with risk. It's easy to forget them and let them go to seed. I usually try to hoe around select weeds early in the season, harvest one nice batch of full-sized succulent tips, then weed them out the next time through the area with a hoe.

Edible weeds are true weeds with respect to having the full repertoire of seed dormancy mechanisms. This means that they can contaminate the soil seed bank for years from a single year's seeding. An edible weed is still a weed.

CHAPTER SEVEN

Beginning—Tomatoes

Begin with Something You Really Love. Tomato Kinds and Colors.
Flavor Favorites. Thirty Interesting Open-Pollinated Tomato Varieties.
Starting Tomatoes from Seed—Growing Transplants. Potting Soil for
Germinating Seeds and Starting Transplants. Preparing the Ground. Hardening
Off and Planting Transplants. Do Carrots Really Love Tomatoes?—Garden
Woman Adventures. Polycultures. Supporting and Nurturing. Watering and
Mulching. Why It Will Soon Be Impossible to Grow Our Current Generation
of Heirloom Tomatoes and What to Do About It—Late Blight 101. Dealing
with Late Blight. Late Blight Resistant Hybrid Tomato Varieties. Late Blight
Resistant Heirloom and Open-Pollinated Varieties. Why the Best-Flavored
Tomato May Not Be the One That Is Picked Vine-Ripe. Using Green Tomatoes.

Begin with Something You Really Love

Beginning gardeners are often encouraged to grow certain crops just because they are so easy to grow. Many vegetable gardeners plant little round red radishes year after year until they suddenly realize they don't like little round red radishes all that much. I managed to avoid getting seduced by the easiness of radishes, but somehow succumbed to chard. Chard is a beet bred for the stems and leaves. It's even easier to grow than radishes. The varieties with red, multicolored, or yellow stems are beautiful. Chard is edible, but I don't know anyone who actually enjoys eating it very much. It's really bad raw. Even cooked, any dish containing chard would taste better without it. From my point of view, the main value of chard is for comparison. Whenever I'm trying a new green and it fails the flavor test, I always tell myself, "Well . . . at least it doesn't taste as bad as chard."

Growing something just because it is easy makes about as much sense as climbing a mountain just because it is there. Would I jump over a cliff just because it was there? Would I throw myself into a blazing inferno just because it was there? Would I try to swim across the Pacific Ocean just because it is there? Certainly not! Mere existence of an entity does not create a requirement for me to interact with said entity in any particular fashion. There are many mountains within sight of my garden, and I have managed to leave all but four or five of them peacefully unclimbed. Likewise, these days I don't plant little round red radishes or chard, however easy they are. You don't have to either.

If you are a beginner, in most cases it's perfectly obvious where to begin vegetable gardening. Above all you want homegrown tomatoes. If you are an experienced gardener, homegrown tomatoes are usually still one of the top priorities. You want more of them, and you want them earlier, and you want the most delicious varieties, and you want different kinds and colors.

Tomato Kinds and Colors

According to the USDA, tomatoes are not actually an essential food group. Many of us gardeners would disagree. Tomatoes contain substances we perceive as the flavor umami, or savory, more usually associated with meat and meat broths instead of vegetables. Umami is one of the five basic tastes in addition to sweet, sour, salty, and bitter. Umami flavor rounds out and enhances other flavors, particularly in the presence of salt. Other potent sources of umami flavor besides meat are fish, cheese, mushrooms, and fermented products of many kinds such as soy sauce, tamari sauce, and miso. Umami flavor often involves glutamate, the salt form of the amino acid glutamic acid. Tomatoes are high in glutamate. However, there are other compounds including various nucleotides (nucleic acid components and breakdown products) that also give an umami flavor. The richest, roundest umami flavor comes with a combination of multiple umami-receptor-triggering substances. This is undoubtedly why we are so inclined to combine tomatoes with meat or cheese or both. Curing meat creates and enhances umami flavor. If we indulge in bacon, lettuce, and tomato sandwiches, or hot dogs with sauerkraut and ketchup, or pepperoni pizza, or even bacon cheeseburgers with ketchup, let no one criticize us. We are just striving for the ultimate umami experience.

For most gardeners, tomatoes are the starting point and main issue. Any year in which we grow as many tomatoes as we want is a good year. But which kinds of tomatoes do we want to grow?

Lots of different words get used in talking about tomato varieties. First, *different tomato varieties can be determinate or indeterminate.* Determinate tomatoes grow as bushy plants that deliver their harvest of ripe tomatoes in a relatively narrow time frame. This is ideal for processing, or for the first early tomato crop to hold us until our main-crop varieties mature. Determinate tomato plants are often grown without supports or with minimal support such as the wire ring structures available commercially. Indeterminate tomatoes have a vining form and usually give you a much greater harvest of tomatoes for the amount of space involved. They usually keep bearing tomatoes until the end of the season. They ordinarily require serious support. They usually are spaced wider than determinate types. Some tomato varieties compromise; they have a bushy form but continue growing and bearing over a longer season than complete determinates. These types are sometimes called *compact indeterminates* and sometimes are lumped with determinates. Basically, you grow them as determinates.

The tomato varieties that win the flavor contests are almost always indeterminate varieties. They have a bigger ratio of leaves to fruit, so are more able to load sugar and flavor into their fruits. When we grow determinates, we usually are settling for a bit less flavor in exchange for other practical virtues.

When you buy transplants, they should have the name of the variety as well as whether the variety is determinate or indeterminate. If you have a small garden where getting the most out of each square foot is essential, you focus on indeterminate varieties for your main crop, and use determinate varieties, if at all, only for the first early crop. Most of the very earliest varieties are determinates. In addition, you can make a low hoophouse for determinate varieties early in the season and bring them into bearing even faster. It's harder to give protection to the bigger, more rampant-growing indeterminates.

Tomatoes can be juicy tomatoes or they can be slicers or paste types. The 'Stupice' and 'Pruden's Purple' I grow are juicy types while the 'Amish Paste' is a paste type. Juicy tomatoes are great for eating out of hand or using in various dishes from soups to salads. However, they aren't ideal for using in sandwiches because they are so wet they make the bread soggy, that is, if all the insides don't drip out before you even get them to the sandwich. The slicing tomatoes sometimes have more solid flesh and less juice. In other cases the amount of juice is the same as would be expected for a tomato of that size, but the seed cavities are more numerous and smaller so their contents don't fall out when the tomato is cut. However, there is considerable overlap and ambiguity between slicers and juicy types. I consider 'Brandywine' and 'Pruden's Purple' to be juicy types, but they are sometimes referred to as beefsteak types or slicers.

Paste tomato varieties have the most solid portion and the least juice of all tomato varieties. They usually don't taste all that great as raw eating tomatoes, though there is at least one noteworthy exception ('Amish Paste'). A paste tomato is designed to taste great after the tomato has gone through cooking into canned tomatoes or ketchup or spaghetti sauce. Paste tomatoes have very few seeds per tomato because of the tiny seed cavities. This minimizes waste in processing where seeds are removed and makes it easier to not care about the seeds if you just leave them in. Many (but not

all) paste tomatoes have a special shape, an oblong shape with the scar where the tomato attached to the plant being an "outie" rather than an "innie" in configuration. This distinction is important in processing time. It takes several thrusts of a knife to remove the stem scar from a tomato of ordinary shape. With an oblong tomato with an outie stem scar you just slice off the stem-scar end with a single cut.

Tomatoes can be various colors other than red when fully ripe. Yellow, orange, pink, purple, brown, black, green, and various patterns of these colors in stripes or whorls are all possibilities. The different colors are usually associated with different flavor classes that taste distinctly different from red tomatoes. Ordinary red tomatoes have red flesh and yellow skins. Pink tomatoes have red flesh or pink flesh and clear skins. (Some look as red as red tomatoes.) Green tomatoes are tomatoes with a gene that causes them to fail to lose their chlorophyll during fruit ripening stacked on top of another gene that causes the tomato to fail to produce lycopene, the red color. If the tomato keeps its chlorophyll but also produces lycopene, we get a "black" tomato. Black tomatoes are not really black, but a mixture of shades of green, black, red, brown, and purple where different amounts of green and red pigments mix together. There are also some really black tomatoes that are genetically overproducers of anthocyanin in their skins (such as 'Indigo Rose').

Red tomatoes have the advantage that it is very easy to tell when they are ripe. With the other colors, it takes more time and in some cases experience with the particular variety. This is especially true for the blacks and even more so for the greens.

Recently many seed companies have started to sell various *tomato varieties grafted onto hybrid rootstocks.* They claim the grafted tomatoes are highly vigorous, disease-resistant, faster than a speeding bullet, and able to leap tall buildings in a single bound. Okay, not quite, but nearly. I haven't tried grafted tomatoes, and can't tell you whether they yield better or not. What I can tell you is if we let big seed companies convince us that we need to have even our heirloom tomatoes grafted onto proprietary rootstocks in order to grow homegrown tomatoes, it will be a major step toward making us completely dependent on these big seed companies. We will not be able to use our own saved seed, but will have to buy tomato plants every year and limit ourselves to only the varieties the companies that do the grafting choose to make available to us. I would rather be able to save seed myself and start tomatoes myself, or buy plants of the varieties I care about from small local farmers. Producing tomatoes by using varieties grafted onto proprietary rootstocks is a direction I don't want to go in or encourage.

Some recent tomato breeding has focused on certain *nutritional characteristics.* I can't say I worry much about it. Lycopene, the red pigment in red tomatoes, is an antioxidant, but so are lots of substances in foods. There is no reason to think that our health is in danger because we are growing a tomato variety with less lycopene than some other variety. Likewise for carotene. None of the high-carotene tomatoes I have tried measures up to my favorite reds or pinks in earliness or flavor. These might matter a lot if they were used to make ketchup and you are someone for whom ketchup is your only vegetable. (And, sadly, there are people for whom that is the case.) However, do *I* need more carotene when I eat so many greens and golden summer squash and orange-fleshed winter squash? So much so that I should choose my tomato varieties based on their having a higher content of whatever single compound is some breeder's favorite nutrient du jour? I pass. I eat a big variety of colorful foods and undoubtedly get all the antioxidants I need.

Actually, it's only an assumption that consuming as many antioxidants as possible is good for us—or, for that matter, that free radicals in all amounts are bad for us. There is already one study showing otherwise. It's been shown that if we take large doses of antioxidants to counter the free radicals produced by a strenuous exercise program, the antioxidants actually prevent the conditioning effect of the exercise. The free radicals produced when we exercise are apparently an essential part of exercise-induced conditioning.

I choose my tomato varieties according to what grows and produces and tastes best. These are usually heirlooms. However, my friend Alan Kapuler speculates that heirlooms might be more nutritious than most varieties, whether or not in ways we currently understand. Heirlooms are, after all, by definition the varieties whose caretakers survived long enough and well enough to have families to pass their seed on to.

Tomato varieties can be hybrid or open-pollinated varieties. Hybrids don't breed true when saved from seeds. You have to buy the seed from a seed company every year. Open-pollinated varieties breed true, and we can save their seed. *Heirloom tomatoes*, like other heirloom vegetable varieties, are open-pollinated varieties that have been around for a while and have been passed down and cherished like other heirloom possessions. Like other OP varieties, they breed true when you save the seed. Certain heirloom varieties almost always win the flavor taste tests. Commercial varieties meant for market gardeners have to meet many criteria, and flavor is not at the top of the list. We home gardeners have the luxury of putting flavor first. There are plenty of spectacularly delicious heirloom varieties that grow and yield so well that a little more or less yield compared with a bland-flavored hybrid is not even noticeable, let alone being the main issue.

Heirloom tomatoes often have *greenish or off-color shoulders* and areas around the stem scar that are not quite ripe when the rest of the tomato is prime. Most modern breeders have considered this characteristic a devastatingly serious defect and have bred strenuously for *uniform ripening* by making use of the recessive gene *u*. This gene is bred into nearly all the commercial varieties and hybrids. It causes the tomato fruit to lose much of its chlorophyll as it goes into the ripening stage. Fruits of plants that are homozygous *uu* are light uniform green before ripening, then ripen into fruits without green shoulders. However, a recent study shows that *u* also causes the fruits to lose a good portion of their sugar and flavor. Apparently tomato fruits need the full amount of chlorophyll in the fruit itself while ripening in order to develop full flavor. So if it seems you always get the best tomato flavor from the irregular "ugly" heirloom with green shoulders instead of from the visually perfect modern varieties, it's because you do. Just chop the small unripe portion off your prime heirloom tomatoes and give thanks for life at its richest, the sublime along with the imperfection.

Hybrid tomatoes are not usually the choice of a home gardener who puts flavor first. Some hybrids, particularly some of the older ones, have good flavor, though never in quite the same league as the flavor of the best heirlooms. However, most of the more recent hybrids have been bred primarily for visual appeal, and contain the flavor-killing, sugar-killing gene for uniform ripening. Some hybrids may yield better than most OP varieties, but not usually enough to be noticeable to a home gardener, and not by enough to make up for the better flavor of the best heirloom and OP varieties, let alone for the lack of control over the seed.

Hybrids and some modern commercial OP varieties are sometimes touted as being *crack-resistant*. I grew one of the recent hybrids last year, and I can confirm that it was indeed crack-resistant. The skin was so thick and tough I could hardly get my teeth into it. The fruits were uniformly red-colored and looked glorious. But they tasted like wet cardboard. The variety also had genes for slow ripening so it would have a long shelf life. It certainly had a long life on *my* shelf. Weeks. I didn't eat the tomatoes because at no point was I totally sure they were food.

Where the modern hybrid tomatoes excel is in being *resistant to many diseases* that matter in commercial plantings. Most of these diseases haven't mattered much in traditional home gardens as long as we rotated the land we grew the tomatoes on from year to year. However, late blight is changing this picture. I consider the spreading of new virulent lines of late blight and its implications in separate sections.

Flavor Favorites

Tomato flavor is very much affected by growing conditions. A variety that tastes great grown in one place might taste inferior when grown elsewhere. And not all varieties do well in all locations. I'm going to start by telling you about my favorites, which I hope will help you realize the kinds of factors that might matter, and assist you in making your own choices. Then I'll mention some other favorites of those gardeners in other regions that have more summer heat than we do in the Northwest.

Here in the Willamette Valley of Oregon we have "mild" summers, which sounds great until you restate that as "we have very little summer heat, too little to grow and ripen the longer-season tomato varieties." So we generally grow early or midseason varieties. In addition, some varieties we can grow that are famous for flavor elsewhere taste bland here. It apparently takes more summer heat to realize the full flavor potential of some varieties.

My standbys, the tomatoes I grow every year in good amounts, are 'Amish Paste—Kapuler', 'Stupice', and 'Pruden's Purple'. All of these grow well here, produce tomatoes of excellent flavor even in our colder summers, and can even produce good-flavored tomatoes in a somewhat shady spot. Two others that have more recently made my list of regulars are 'Black Krim' and 'Legend'. The tomato at the top of my list to try next year is 'Geranium Kiss', a new open-pollinated tomato bred by Alan Kapuler and Kusra Kapuler that has turned out to be highly late blight resistant.

'Amish Paste—Kapuler' is my main-crop tomato. This line of this heirloom variety, selected and preserved by the Kapuler family for decades, is the best line of 'Amish Paste'. Other lines may be later, less vigorous or productive, less tolerant of cool weather, more prone to cracking, or seedier. 'Amish Paste—Kapuler' has much bigger tomatoes than most paste varieties, up to about 12 ounces (340 g). The tomatoes on each plant are a bit variable in shape, with some

long and relatively thin and pointed like Roma types and some almost as fat as they are long and more teardrop- or oxheart-shaped. Unlike most paste tomatoes, 'Amish' has excellent flavor raw, so can be used as a general-purpose eating tomato as well as for processing. 'Amish' is often among the top-rated ones for flavor in trials of fresh-eating tomatoes. It also usually beats better-known paste tomatoes such as 'San Marzano' and 'Roma' when trialed for making tomato sauce. 'Amish Paste' has even been designated a Slow Food Ark of Taste variety. I think it would win an even larger proportion of the contests in all categories if more people were using the Kapuler line instead of various other lines. Because 'Amish Paste—Kapuler' plants are big and vigorous, I space them at 4 feet (1.2 m) apart in the rows and give them sturdy support.

In her book *100 Heirloom Tomatoes for the American Garden*, Carolyn Male says "'Amish Paste' . . . is actually quite juicy and has an average amount of seed." She also describes the plant as wimpy and wispy with little leaf coverage. Not the Kapuler line. The flesh isn't dry, but the fruit is almost all meat with little free juice, and seed production is *very* low. And the plants are vigorous and provide plenty of leaf cover. The website tatianastomatobase.com, a wonderful source of information on tomato varieties, gives a description similar to that of Carolyn Male, and a size for the tomato of 4 to 8 ounces (113 to 227 g). This is laughably small to those of us used to the Kapuler line of 'Amish Paste'. There are apparently plenty of inferior lines of 'Amish Paste' out there. With many varieties, the specific line of the variety really matters, and the differences between lines can be as large as or larger than the differences between many varieties.

I use 'Amish Paste—Kapuler' not just for fresh eating but also to make tomato juice, which I freeze and use to jazz up spaghetti sauce, tomato soup, and other tomato-based dishes in winter. There are so few seeds I just leave them in, making for very easy processing. The juice is spectacularly flavorful. 'Amish Paste' also makes wonderful dried slices. I just slice the tomatoes into slices about ⅜ inch (1 cm) thick and dry them in a dehydrator at 125°F/52°C.

I sell 'Amish Paste—Kapuler' through my own seed company, Fertile Valley Seeds. Others I know who sell this line are Nichols Garden Nursery and, of course, the two Kapuler family seed companies, Peace Seeds and Peace Seedlings.

'**Stupice**' is the tomato I use for my first early crop. It's a smaller-vined indeterminate variety from Czechoslovakia. It is consistently earlier than everything else I have tried with the exception of extreme early determinates with poor flavor. The tomatoes of 'Stupice' are only up to about 2.5 inches (6.4 cm), which is smaller than I like. However, it is, as far as I know, the earliest tomato with excellent full-season tomato flavor. 'Stupice' is a juicy type. It's a lovely size for eating out of hand. In the early part of the season I use Stups for everything. I used to plant just a couple of 'Stupice' plants for my first tomatoes of the season, then forget them when the bigger tomatoes came on. With our increasingly cold summers and increasingly later tomato harvest, I've started planting more 'Stupice' and depending on them for a larger part of the season. Because 'Stupice' is a smaller-vined indeterminate I space plants 2 feet (0.6 m)

apart in the rows instead of the 3 to 4 feet (0.9 to 1.2 m) I use for bigger indeterminates. ('Stupice' is a genuine vine type, just smaller, not an overgrown bush.) There are lots of cherry tomatoes that taste good and do well here, incidentally. I don't grow them mostly because I have found picking them to be too much work, and none with good flavor are earlier than 'Stupice', which has much bigger tomatoes than cherry tomatoes. 'Stupice' is widely available.

'**Pruden's Purple**' is my pink tomato of choice. It is similar in flavor to 'Brandywine' but ripens two or three weeks earlier. The tomatoes are also huge for such an early plant. It's usually listed with a maturity of sixty-seven days with tomatoes up to a pound (453 g). Some people prefer the flavor of 'Pruden's Purple' to 'Brandywine', others vice versa.

Red tomatoes have a yellowish skin and red flesh. Pink tomatoes have red flesh but a clear skin and a flavor different from the reds. As with other red-fleshed tomatoes, the shade of red varies depending on the variety. 'Pruden's Purple' has very red flesh, and the whole tomatoes don't look pink, they look red. However, they carry the gene for clear skin and have that distinctive pink tomato flavor in full measure. (Some but not all varieties with *purple* in their name belong to the pink class and have pink tomato flavor. Some varieties with *purple* in the name are actually blacks with a quite different flavor. 'Pruden's Purple' is a pink. But 'Cherokee Purple' is a black, for example.)

'Pruden's Purple' is widely available.

'**Black Krim**' is the black tomato I grow. It is a little later than 'Amish' and 'Pruden's'. The exterior and interior are a strange-looking mix of purple, red, black, brown, and green. Black tomatoes retain their chlorophyll during the ripening stage so you get green and red coloring mixing together in various proportions in the ripe tomatoes. (Black tomatoes may have either yellow or clear skin.) Traditional black tomatoes are more brown or mahogany with just a little black rather than actually black. Tomatoes that are genuinely black, such as the recently released 'Indigo Rose', are genetically quite different.

'Black Krim' tomatoes are big, often bigger than 1 pound (453 gm). 'Black Krim' has a flavor similar to the best reds but with components reminiscent of barbecued meat fresh from the grill accompanied by a full-bodied red wine. It's a wonderful flavor. Chunks of Krims are especially spectacular dropped into beef soup or stew just before serving. Krims are also my favorite tomato for dressing with hot spaghetti sauce and serving on rice or polenta.

'Black Krim' tomatoes are fully ripe when they are about half green and half purple/brown on the outside and still quite firm. If you wait until the fruits are purple/black all over or until they are soft, they will be so far past prime that they actually taste somewhat rotten. This variety in the farmers markets is usually picked too late and usually has overtones of rot, a fact that kept me from trying the variety in my own garden until I read about its ripening characteristics in the Fedco catalog. It's harder to tell when the blacks are ripe; you have to test and learn with each black variety.

'**Legend**' isn't on my flavor favorite list, but I have started growing it regularly. It produces toma-

toes with acceptable flavor for such an early tomato and has other important virtues. It's a vigorous determinate bush that produces early and produces huge tomatoes, sometimes up to a pound. It is about a week later than 'Stupice' but makes huge tomatoes rather than small ones. With our summers being so cold recently and our outdoor tomatoes so late, I plan to start planting more determinates for the first early tomato crop and just covering them with a low plastic tunnel during the first part of the season. Furthermore, I will plant such protected determinates earlier, at the average last frost date instead of safely after it. These two tricks should give me an extra month of tomatoes, I figure.

'Legend' is huge for such an early tomato. It's usually listed as sixty-eight days, and fruits weigh up to 1 pound (453 gm). In addition, it is parthenocarpic, that is, it can produce fruits without pollination. This means the plant can set fruit in weather too cold for the release and function of the pollen. In addition, tomatoes grown in low plastic tunnels out of the wind may have to be uncovered and jostled regularly to set fruits. This won't be necessary with a parthenocarpic variety.

Finally, 'Legend' is homozygous for the late blight resistance gene *Ph2*, and is highly tolerant to late blight strains US-8 and US-11. It also proved somewhat but not adequately resistant to US-23 in tests on Long Island in 2012. Generally, resistance is a matter of the specific lines of blight that are available from year to year and region to region. I suggest growing 'Legend' not just for very early big tomatoes, and as backup in case there is blight, but also to use in breeding our own blight-resistant tomatoes as I describe in the last chapter.

'Legend', bred by Oregon State University plant breeder Jim Baggett (now retired), is an open-pollinated public domain variety with useful amounts of blight resistance for many situations. The fact that 'Legend' is parthenocarpic means that you can have lots of fruits with no seeds, a characteristic we seed savers can find a bit inconvenient. Fortunately, the late fruits produced in warm weather usually have seeds. In crosses 'Legend' can contribute not just blight resistance but also earliness, size, vigor, the ability to thrive in cool climates, and parthenocarpy (if you want it).

Nichols Garden Nursery is the best source of seed for 'Legend'. It's easy to accidentally select for more seeds in a parthenocarpic variety, that is, for accidental crosses or for mutations that change the variety. Rose Marie Nichols McGee tells me they go back to OSU material frequently for starting seed and contract for their own growouts, ensuring that their 'Legend' stays true to the original.

'Geranium Kiss' is a new variety bred by Alan Kapuler and his daughter Kusra Kapuler. It produces stocky determinate plants with hypertresses of twenty to seventy 1- to 2-ounce (28 to 56 g) fruits. The plant makes three to four settings of hypertresses. The foliage is dark green and looks like a geranium. What was most interesting to me, however, was the final line in the Peace Seeds catalog listing: "When a long cool rain comes in the fall, most of the tomato plants get a late blight. This one does not." Knowing Mushroom (Alan Kapuler) does a lot of breeding with wild tomatoes, I immediately emailed back and forth with him a bit to get further information. If this blight resistance came in from a wild tomato, it might be different genetically from the blight resistance the university breeders are playing with, so could be invaluable.

The original cross that was bred and stabilized to create 'Geranium Kiss' did, indeed, involve a wild species. It was a cross of 'Stakeless' with the grape tomato, *Solanum humboldtii*. Mushroom (A. K.) thinks the late blight resistance came from *S. humboldtii*.

"In the field it is quite striking when a fall rain comes and most of the tomatoes are decrepit while the 'Geranium Kiss' is robust and healthy," Mushroom says. "*S. pimpinellifolium*, the current tomato, got knocked down by the blight (i.e., 10 inches [25 cm] of rain in September) but not as bad as all the other tomatoes except 'Geranium Kiss'."

The current crop of late blight resistance genes (*Ph2* and *Ph3*) all come from *S. pimpinellifolium*. The *humboldtii*-derived resistance in 'Geranium Kiss' might be entirely different.

'Geranium Kiss' is available from Peace Seeds, Peace Seedlings, and Nichols Garden Nursery.

Nichols Garden Nursery and Territorial Seeds have extensive lists that include all the tomatoes most commonly grown in the maritime Northwest. For the most extensive list of rare varieties that grow well in the maritime Northwest, see the Adaptive Seeds website and catalog. Adaptive Seeds is the seed company branch of The Seed Ambassador Project, and their list is based heavily on their own independent germplasm-collecting expeditions to organic and biodynamic farms in Europe. Most of the earlier (but not the later) tomato varieties that do well in New England also do well in the maritime Northwest, so we in the

Northwest also regularly go to Johnny's Selected Seeds, Fedco Seeds, and High Mowing Organic Seeds for tomato and other varieties.

To get an idea of what tomato varieties you can grow that have the most spectacular flavors if you have a bit more summer heat than I do, I consulted CR Lawn, of Fedco Seeds, who is as passionate about tomato flavor as I am. He gardens in Massachusetts.

Here is CR's list of flavor favorites, in order of preference: 1. 'Pink Brandywine'. 2. 'Aunt Ruby's German Green'. 3. 'Cherokee Purple'. 4. 'Cosmonaut Volkov'. 5. 'Lillian's Yellow Heirloom'. 6. 'Rose de Berne'. 7. 'Goldie'. 8. 'Tiffen Mennonite' or 'Weisnicht's Ukrainian'.

CR's favorites in cherry tomatoes are 'Sungold Hybrid' and 'Honeydrop'.

When it comes to flavor and extra-early tomatoes, CR says "'Glacier' is the best of the very earlies, but cannot compare with any of the nine above." 'Glacier' is actually reported to be able to tolerate light frost.

The top three varieties on CR's list are all frequently on lists for top flavor, are widely available, and are big tomatoes with classical tomato shape.

'Cosmonaut Volkov' is the only determinate tomato on CR's flavor favorite list, and it is very early (sixty-five days). I'm going to have to try it. As a rule, determinate tomatoes don't win flavor contests. 'Cosmonaut Volkov' seems to be breaking that rule. Judging from the pictures I've seen of 'Cosmonaut Volkov' with its greenish shoulders, it does not carry the flavor-killing *u* gene, the gene for uniform ripening that has been bred into so many of the North American determinate varieties.

Those who live farther south or farther inland who have more total summer heat can grow a much wider range of tomato varieties than those of us in the maritime Northwest or New England. This doesn't mean you can grow everything those in cooler areas can grow, however. To thrive in Virginia or Missouri, a tomato needs heat resistance and an ability to handle warm nights, not an ability to produce in cool weather and tolerance to cold nights. And it may also need more or different kinds of disease resistance. Some varieties are very widely adapted; others are best only in certain regions.

Southern Exposure Seed Exchange, based in Virginia, is a regional seed company providing heirloom and open-pollinated varieties for the mid-Atlantic and Southeast. Their catalog lists more than a hundred tomato varieties, most carrying the special sun symbol designating varieties that do especially well in their territory. Ira Wallace, who works with Southern Exposure, is author of the recently released book *The Timber Press Guide to Vegetable Gardening in the Southeast*. She writes that in the test gardens and at the public taste testings that Southern Exposure sponsors, the following varieties are consistent favorites:

For cherry tomatoes, 'Matt's Wild Cherry' (sixty days) and 'Amy's Apricot' (seventy days) are the favorites. 'Matt's Wild Cherry' produces lots of tiny tomatoes with intense flavor. 'Amy's Apricot' "is a flavor contender for an OP alternative to the ever-popular hybrid 'Sungold'."

For red slicers Ira suggests 'Old Virginia' (eighty days) and 'Tropic VFN' (eighty days).

For paste tomatoes she suggests 'Roma Virginia Select' (seventy-five days), a farmer-selected variety, and 'Amish Paste', the largest paste tomato she has grown.

For the special colors, Ira suggests 'Garden Peach' (seventy-two days) (yellow), 'Cherokee Purple' (eighty-five days) (a black), and 'Eva Purple Ball' (seventy-eight days) (a pink). "'Eva Purple Ball' is one of my all-time favorites . . . the color could be described as a deep rose," Ira says.

Finally, for the really big heirloom slicers Ira suggests 'Granny Cantrell's German Red/Pink' (sixty-nine days), a tomato with a flavor similar to 'Brandywine', and 'Mortgage Lifter VFN' (eighty-three days) (red).

Southern Exposure sells all these varieties.

If you live in the Midwest you have enough summer heat to grow many southern-adapted as well as many northern-adapted varieties. The main consideration seems to be whether your growing season is long enough for the variety. In the Thirty Interesting Tomatoes section below I list the varieties by maturity times in New England. With more summer heat the maturity times will be shorter; however, the order of maturity usually holds true. If you live in the hot, arid Southwest, check out the tomatoes listed in the Native Seeds/SEARCH catalog, which is regional for the Southwest.

Bountiful Gardens, in California, has an excellent and interesting collection of tomatoes for the West, including a special section for those in the Northwest. (They also have impressive collections of grains, including heirloom wheats, amaranths, and quinoas.)

In addition to the regionally oriented seed companies I've already mentioned, these additional companies also have particularly noteworthy lists of open-pollinated tomato varieties: Totally Tomatoes, Trade Winds Fruit, Baker Creek Heirloom Seeds.

A glorious source of information about tomato varieties is tatianastomatobase.com.

Thirty Interesting Open-Pollinated Tomato Varieties

Here is a list showing maturity times from transplanting, plant form, and fruit size and color for thirty interesting varieties. All are open-pollinated. The tomatoes are listed in order of ripening (in New England, unless stated otherwise; at Southern Exposure Seed Exchange in Virginia where stated).

'Glacier'. 55 days. DET Red. Sweden. Frost-resistant. Up to 2.5 inches/6.5 cm.

'Stupice'. 60 days. Small-vined IND Red. Up to 2.5 inches/6.5 cm.

'Matt's Wild Cherry'. 60 days. IND Red. Up to 5/8 inch/1.6 cm. Resistant to late blight; genetic basis unknown. Also some resistance to early blight.

'Geranium Kiss'. Early DET Red. Up to 2 inches/5.1 cm. Dark-green foliage looks like a geranium. Late blight resistant; genetic basis unknown. Resistance gene(s) probably derived from the wild species *S. humboldtii*. Bred by Alan Kapuler and Kusra Kapuler.

'Taxi'. 64 days. DET Yellow. Up to 6 ounces/ 171 g. Popular in Northwest.

'Cosmonaut Volkov'. 65 days. Semi-DET Red. Up to 12 ounces/340 g.

'Legend'. 68 days. DET Red. Parthenocarpic. Up to 5 inches/12.7 cm; up to 1 pound/453 g. Late blight resistant. Homozygous for *Ph2*.

'Amish Paste—Kapuler Line'. 72 days. IND Red. Up to 12 ounces/340 g.

'**Pruden's Purple**'. 67–72 days. IND Pink flavor class; they look red. Many over 1 pound/453 g.

'**Cherokee Green**'. 72 days. IND Green. 8-plus ounces/237-plus g.

'**Granny Cantrell's German Red/Pink**'. 69 days at SESE/Virginia. IND Pink. 1 pound/453 g or more.

'**Black Prince**'. 74 days. IND Black. Up to 3 inches/7.6 cm.

'**Cherokee Purple**'. 72–77 days. IND Black. Up to 13 ounces/369 g.

'**Green Zebra**'. 75 days. IND Bicolor striped green and yellow. Up to 5 ounces/142 g.

'**Goldie**' (aka 'Dixie Golden Giant'). 75 days. IND Deep orange. Up to 20 ounces/567 g.

'**Great White**'. 75 days. IND Yellow. 12-plus ounces/340-plus g.

'**Paul Robeson**'. 78 days. IND Black. Up to 12 ounces/340 g.

'**Striped German**'. 78 days. IND Bicolor yellow and red striped. 12 ounces/340 g.

'**Eva Purple Ball**'. 78 days at SESE/Virginia. IND Pink class (deep-rose color). Up to 7 ounces/198 g. From SESE catalog: "Outstanding performer in hot, humid areas. Excellent resistance to foliar and fruit disease . . . Fruits are easy to harvest, some dropping from the vine at peak ripeness. A wonderful all-purpose tomato with excellent flavor."

'**Old Virginia**'. 80 days at SESE/Virginia. IND Red. Heat- and bacterial-wilt-tolerant. 4 ounces/113 g.

'**Black Krim**'. 80 days. IND Black. Up to 18 ounces/510 g.

'**Rose de Berne**'. 80 days. IND Pink. 5 ounces/142 g.

'**Brandywine**'. 82 days. IND Pink. Up to 1.5 pounds/680 g.

'**Amish Paste**' (most lines, not the Kapuler line). 78–85 days. IND Red. Up to 12 ounces/340 g. Sometimes listed as 4–8 ounces. Lots of inferior lines out there. I suggest getting the Kapuler line.

'**West Virginia 63**' (aka 'Centennial'). 70 days at SESE/Virginia. IND Red. Up to 8 ounces/267 g; 3 inches/7.6 cm. Bred by Mannon Gallegly. Late blight resistant; homozygous for *Ph2*. Also resistant to *Verticillium* and *Fusarium*.

'**Aunt Ruby's German Green**'. 80–85 days. IND Green. Up to 1 pound/453 g.

'**Weisnicht's Ukrainian**'. 85 days. DET (compact IND) Pink. Up to 12 ounces/340 g.

'**Tiffen Mennonite**'. 86 days. IND Pink. 15 ounces/425 g.

'**Lillian's Yellow**'. 88 days. IND Yellow. 1 pound/453 g.

'**Pineapple**'. 90 days. IND Bicolor striped and marbled red and yellow. More than 1 pound/453 g.

Starting Tomatoes from Seed—
Growing Transplants

Tomatoes are tropical plants. They are sensitive to frost and don't thrive in cool weather. To grow tomatoes in temperate climates we honor their essential tropical nature for the first part of their lives by starting and growing them indoors or in greenhouses. We transplant into our gardens only after the weather warms up.

Tomato transplants are widely available in gardening stores and nurseries. If you are a beginning gardener it's easiest to start off with bought transplants initially. However, in order to be able to grow tomatoes of the most delicious heirloom varieties you often have to grow your own transplants. If you want to grow anything unusual, you need to start from seed. If you save seed or want to do a little breeding, you need to be able to start from seed. In addition, when you start your own transplants you don't have to worry about importing diseases with them. You can also produce better transplants. Commercial transplant growers usually produce transplants that sell best, and most buyers prefer larger transplants. However, larger transplants are usually too big for their pots and are already pot-bound. Some even already have flowers and fruit, which is a bad sign, not a good one. Pot-bound transplants suffer a setback when transplanted. When you grow your own you can produce smaller, less impressive-looking, but better-performing transplants.

Most people recommend starting your tomato seeds six to twelve weeks before you want to set the plants outdoors, usually after all danger of frost is over. Since I think small transplants are better, I suggest closer to five weeks unless your growing area is cool, in which case the plants might need longer. If you plan to cover the plants with a plastic tunnel or a floating row cover to provide a little frost protection, you can set them out a bit earlier.

You start tomato seed indoors pretty much the same way you start any other seed indoors. Some people just put seed starting mix or soil in pots with holes in the bottom, poke the seed in, and set the pots in a window. That can work fine if your home is warm enough and the light from the window is bright enough. My home isn't warm enough to start most seed well in the winter with such casual methods. In addition, winter days in maritime Oregon are usually heavily overcast, and the light from a window isn't bright enough. The seedlings grow poorly, reach desperately for the sky, look pale and sun-starved and unhappy, and become leggy, that is, develop tall thin weak stems. I use tricks to provide a little extra heat during the germination stage as well as fluorescent shop lights in addition to the window to provide adequate light.

You can buy special bulbs for growing plants that produce a larger proportion of light in the wavelengths most useful to the plants. You can even buy special fixtures. I just use secondhand shop lights and whatever bulbs they come with or that are handy when it's time to replace them. I prop the light fixture up with bricks so that it is just a few inches above the transplants. I add more bricks to adjust the height as the transplants grow. (Keep in mind that there is an inverse square relationship between light intensity and distance from the source. At twice the distance you get only a quarter as much light. So keep your light as close to the plants as possible.)

For starting tomato seeds indoors I like Speedling polystyrene trays, reusable plastic trays with lots of pyramid-shaped individual chambers, each of which has a hole in the bottom. You can also just use trays of seed starting mix or compost and broadcast the seeds. I like the Speedling trays because it's easier to transplant the seedlings out of them. In addition the trays let me simultaneously start and keep track of dozens of different varieties of tomatoes in one tray.

For pots to transfer the seedlings to once they are a bit bigger I like plastic pots with holes in the bottom that fit into a plastic mesh carrier of about the same dimensions as the Speedling trays. I can put either the Speedling tray or the mesh tray full of pots into a slightly bigger tray full of water to water from the bottom by capillarity.

I always water my Speedling trays or pots with germinating seed or seedlings from below. Since all the pots necessarily have holes in the bottom for drainage, usually large holes, it's too easy to wash the growing medium out of the pots by watering from above. Watering from above can also flatten tiny plants or rearrange the growing medium and uproot them. In addition, wet stems and foliage promote damping-off and other diseases.

Potting Soil for Germinating Seeds and Starting Transplants

Most people recommend you start seeds in a commercial soil-free seed starting mix that is largely peat moss and vermiculite, possibly also containing perlite and/or coconut-husk-derived material, or some equivalent you mix up yourself. (This is not necessarily the same as commercial potting soil, which is often coarser and may contain many other things.) When I start seeds in Speedling trays I do use a commercial mix of this sort, because the mix holds the root-balls together just right for transferring the plants to bigger containers or directly to the field. For bigger pots, whether for starting seeds or for the baby plants once they are ready for more space, I use mostly my own garden soil combined with a little commercial mix or vermiculite to lighten the soil and enhance its water-holding capacity. Then I top each pot off with a layer of the commercial mix to suppress any weed seeds in my soil.

We are usually advised to never use our own garden soil as a potting or seed starting medium unless we heat it in the oven to sterilize it. However, if some seedling is going to turn up its roots and die because of the microorganisms in my garden soil, I would rather find that out sooner than later. And I view the live microbes in my soil as essential cooperators, not undesirable interlopers.

The injunction against using our own soil or compost as a seed starting or potting mix is so universal, however, that when writing this section I got seized by a sudden wave of doubt. "Am I the only one who is just using mostly her own soil and not sterilizing it?" I asked myself. So I asked my friend Alan Kapuler (aka Mushroom) what he uses. I transplant mostly just tomatoes and peppers. I direct-seed everything else. The Kapulers start thousands of plants of scores or even hundreds of species in their greenhouses every year.

Mushroom uses his own compost mixed with some sand, with maybe a little rock dust and kelp meal added. He doesn't sterilize anything.

Starting mixes and potting soils that are just some combination of peat moss, vermiculite, perlite, or coconut husks provide few nutrients to the seedlings. After the seedlings germinate in such a mix you need to fertilize them. We are usually advised to use a fish emulsion or some other complete soluble fertilizer. For the Speedling trays with the commercial mix, I use mud. That is, I add some garden soil to the water tray that I set the pots in for watering from below. I don't fertilize plants in pots that have mostly garden soil. I want tougher transplants that are used to growing in soil, not tender, succulent transplants that are used to being pampered.

Tomato seeds germinate optimally at 86°F/30°C and acceptably well at temperatures between 75°F/24°C and 90°F/32°C. So says the Johnny's Selected Seeds catalog, which gives graphs of seed germination versus temperature for all the seeds Johnny's sells. Tomato seed germinates very slowly or not at all at temperatures much cooler than that, says the catalog, a fact that I have ascertained independently and repeatedly. My house is never above about 68°F/20°C in the winter. I'm not even all that comfortable myself at above 68°F/20°C. Fortunately, the Johnny's catalog goes on to elaborate, the tomato seedlings, once germinated, grow best at 60°F/16°C to 70°F/21°C. So my growing tomato seedlings and my house and I are fully compatible. It's just the brief germination period that is the problem. I put the starting tray somewhere where it will get a little extra heat while the seeds are germinating. A grow mat that provides bottom heat would work if I could find mine. Instead, I usually use my dehydrator with the heat turned low. A warm spot near a radiator or woodstove will also do. A classical seed starting spot is on top of a refrigerator that has its motor at the top and so provides a little extra bottom heat. Presumably you could jerry-rig a box with a light in it to provide a warmer spot, just as people sometimes do to incubate eggs without an incubator.

Evaporation can both dry out the surface of the growing medium as well as lower the temperature drastically. So preventing evaporation may be as important or more important than providing some extra heat. I cover my trays of germinating seedlings with clear plastic wrap or put the entire tray inside a plastic bag to prevent too much evaporation during the germination period. I check the setup with a thermometer before seeding to make sure I have an acceptable temperature. I also open the wrap or bag every other day or so for a little air exchange; however, I don't know whether this is actually necessary.

To fill the Speedling trays with planting medium, I start by dumping an excess on the tray, smoothing it over the tray, then pressing the dry mix into each chamber and scraping away the excess. Then I water the tray from below by setting it into a bigger tray of water. Some of the chambers have too little medium, and it shows up when the medium is wet. So I top off the underfilled chambers while the Speedling tray is still in the watering tray so all the chambers are filled and their contents moistened. It takes an hour or so for the medium to moisten by capillarity (not minutes). I never press down on the growing medium once it is moistened. (It can squeeze out through the relatively large holes in the bottom of each chamber.)

I plant each tomato seed by making a small slice in the moist growing medium in the Speedling chamber with a kitchen knife. Then I press the seed into one side of the slice at exactly the depth I want using tweezers or forceps. Then I use the knife to close the slice over and around the seed.

I water frequently enough to keep the very top of the soil dry but the rest moist.

When it is time to transplant the started seedlings from Speedling trays to bigger pots, I use a pencil cut off to a flat end and whittled down to dimensions that fit nicely into the holes in the bottom of the Speedling tray chambers as a poker. I lift the tray and use the poker to push each plantling's root-ball with attached growing medium up from below to loosen it.

I never grasp or handle the seedling by its stem. It's easy to damage the stem and interfere with its ability to transport water and nutrients. Instead, I lift each seedling gently by its biggest leaf. It doesn't matter much if you damage one leaf.

(Most of these tricks for growing and handling transplants in Speedling trays are courtesy of Mike Hessel, a local melon grower who developed them and taught them to me. Thanks, Mike!)

Preparing the Ground

Since we plant tomatoes as relatively large transplants and are digging a special hole for each plant anyway, and since tomatoes make vigorous root systems, we don't need a particularly fine seedbed for the tomato patch. I often use tomatoes to break ground. That is, I put them on ground I've just turned over for the first time, or in the spot that the tiller did the worst job on that year. If the soil is very shallowly tilled, I depend on the planting holes dug into it to give the tomatoes a good start. Then I mulch so that the rest of the soil is softening and improving and being worked over by earthworms as the tomato roots expand into it.

Ordinary garden soil or fertilizing regimens for growing the general vegetable garden may work, but sometimes provide too much nitrogen for tomatoes. With too much nitrogen tomato plants grow vegetatively very vigorously but are slower to flower, slower to set fruits, and much slower to ripen their fruits. If I have had a good legume or legume-grass winter cover crop to till under in spring, I don't fertilize my tomato patch at all. Sometimes when I break new ground that hasn't had any legumes in it and plant tomatoes, I fertilize just by mulching gradually with grass clippings. I add just a sprinkling of about half an inch

(1.3 cm) at a time, a thin enough layer so that it dries into grass-clipping hay. When I have about 3 inches (7.6 cm) I stop. That layer acts as both the mulch and the fertilizer. Grass clippings from rapidly growing grass in spring release enough nitrogen and nutrients to provide all the fertility the tomatoes need. (In fact it might provide too much fertility if the soil was good garden soil instead of a new bed. In good garden soil I usually mulch my tomatoes, if I mulch them, with low-nutrient mulches like straw or leaves.)

The nutrition problem people most often run into with tomatoes besides giving them too much nitrogen is giving them too little calcium. Tomatoes seem to need more calcium than many garden plants. If your tomatoes have blossom-end rot—rotten places centered on the blossom scar—it is often a sign that they need more calcium, though sometimes variety and other factors can be involved. Someone once showed me his tomato patch in which virtually every tomato had a rotten end. He was worried his tomatoes had some awful disease. I suggested just sprinkling some agricultural lime, that is ground limestone rock, on the surface of the soil. It completely cured the problem, and the next flush of fruits had no blossom-end rot at all. (I was a bit surprised that enough calcium dissolved from the ground limestone rock just spread on the surface of the soil to fix the problem that fast, but apparently it did.)

Other than providing enough calcium and avoiding too much nitrogen the biggest soil issue for tomatoes is the need for rotating the crops. I'm pretty casual and sloppy about most rotations, but I'm serious about rotating my solanums (nightshades)—that is, tomatoes, potatoes, peppers, eggplants, tomatillos, and ground cherries. I put my potatoes in a patch by themselves since I don't water them. I put all the rest of the solanums together in one area to facilitate rotation. I'm careful not to put the irrigated solanums or the potatoes where I have had either before in the last two years. Most soilborne diseases are not a problem as long as you rotate your solanum crops. If you grow heirloom tomatoes without practicing rotation you are likely to lose a lot of them to soilborne diseases. (Many modern hybrids, however, are resistant to these diseases.)

Hardening Off and Planting Transplants

If you buy your tomato transplants, these days it is especially critical to examine them for disease and to buy only locally raised transplants that have the local farm or nursery name on the tags. Do not buy transplants from big-box gardening stores. Such transplants may come from southern regions where late blight does not die off in the winter. Some occurrences of late blight in recent years in the Northeast are thought to have started because of tomato transplants imported from growers in the South. All it takes is one infected transplant and your garden could be the cause of an epidemic that destroys tons of tomatoes and potatoes in your region. Grow your own transplants or buy locally. And inspect the leaves and stems carefully for the distinctive lesions associated with blight.

Transplants we buy these days often come straight from the greenhouse. I always ask if they have been hardened off, that is, put outdoors for a few days so they adjust to our ordinary outdoor conditions. Usually they have not. It is extra hard on a plant to make the adjustment from greenhouse to outdoors and from pot to soil at the same time. The transplanting usually destroys much of the root system, impairing the plant's ability to absorb even water for a few days until it can grow more roots. So I always set my plants outside in their pots for a few days to let them first adjust to outdoor weather. Then, only after they have made that adjustment do I ask them to deal with the stress of being ripped from their pots and moved.

I water the transplants as needed during the hardening-off period. I prefer to water the plants from below, that is, to set the pots in a tray of water. I always water the transplants thoroughly the evening before the day they are to be transplanted, as they will have trouble absorbing water or anything else for a while after their root systems are disturbed and so many of the root hairs destroyed. It helps if the plants start fully hydrated.

The best day for transplanting is an overcast mild day with no wind and little or no sun. If hot sunny weather is all I have, I transplant in late afternoon so the plants will have at least the night to start establishing a root system that can absorb water. If the conditions are harsh enough to warrant it, I even give the plants some protection from sun. I might, for example, put a large plastic bucket or other opaque object immediately south of each plant so that it is shaded during the hottest part of the day.

Another trick I have for providing enough water for transplants is to water the leaves as well as the soil. Leaves of most plants are quite efficient at absorbing water. Leaves aren't harmed by transplanting.

The morning of the day before T-Day (transplanting day), I cut all the lower leaves and suckers off each transplant, leaving just bare stems with a few leaves at the tops. If there are any flowers

I remove those too. Baby plants will be stunted if they are allowed to produce fruit. Leaves transpire, that is, lose water through their leafy surface. That water loss will put a huge strain on the damaged root system of the plant right after transplanting. So I remove all the lower leaves to bring the amount of leaf surface into balance with the damaged roots. Since I bury the lower part of the stem when transplanting, these leaves would not do the plant any good anyway. Because I make the cuts the morning of the day before, they have time to heal and seal over before I water the plants that evening. If I make the cuts when I water, the cuts will bleed fluid and sap longer instead of scabbing over quickly.

I never handle the stem of a tomato or any other transplant. I loosen the root-ball by poking up from the bottom, or if necessary with some pots, by running a knife around the edge. Then I hold one hand on top of the soil or growing medium around the plant stem with my third and fourth fingers straddling (but not touching) the stem. I turn the entire pot over with the other hand, transferring the plant root-ball from the pot to my hand.

Most transplants are best planted so that the part that was underground in the pot is buried and the rest is above the soil surface. Tomatoes are an exception. They don't mind having their stems buried. The buried stem grows adventitious roots from every node, thus providing a more extensive root system. In addition, the stem of a tomato transplant is often weak and leggy compared with the stem that grows when the plant is outdoors in the air and wind. So I prefer to bury most of the stem.

If I dig a 10-inch (25.4 cm) deep hole so I can bury the stems of my transplants with the plants in an upright position, I have to dig down below the layer of soil that is usually softened by tilling. In addition, in spring the soil is quite cold even in daytime that far down. I prefer to have most of the root-ball and buried stem closer to the surface of the ground where the soil warms up more. So I dig a short trench about 8 inches (20.3 cm) deep and 12 inches (30.5 cm) long and place the transplant in it at a slant with the top with the leaves just above the soil surface and supported by a clod of soil. Some of the roots end up being 8 inches deep, others quite near the surface. There are several buried nodes, and one or more will probably be at just the right level to get enough heat and moisture to establish a vigorous secondary root system.

I mud the transplants in. That is, I dig the holes and fill them with water and let them drain, fill them again and let them drain, then plant the transplant into the mud. I knock any brown or dead-looking roots off the transplant as well as any pot-bound tangled roots. Then I press the root and stem into the mud and surround and support it with mud. Finally, I cover the muddy planting hole over and make a little dike around the area where the root-ball is so that I can easily water just that area initially.

Do Carrots Really Love Tomatoes? —Garden Woman Adventures

Garden Woman reads a popular book on the love between carrots and tomatoes and various other garden plants. It sounds great. Garden Woman delights in the idea of us cooperating and helping one another and getting along. She also delights in the possibility of saving space and work. Garden Woman doesn't worry about the fact that there are no actual data in the book because she knows there are lots of gardening practices without much data behind them, many of which work just fine. So she'll try growing her carrots and tomatoes together and just see.

But exactly how should she do it? If she plants her tomatoes at the same time as the first planting of carrots they will be killed by late-spring frosts. If she waits until the right time for planting tomatoes, she will miss the best and easiest time to grow carrots and the entire early carrot crop. Furthermore, she usually plants her tomatoes in relatively crudely tilled soil, and in single rows with 7 feet (2.1 m) between rows. She usually plants her carrots in beds 3 or 4 feet (0.9 to 1.2 m) wide in soil as finely prepared as she can make it. She ordinarily plants carrots with the fast-growing greens in an area she fertilizes pretty generously, but gives the tomato patch only minimal fertilizer—none if there was a decent winter cover crop. She usually mulches the tomatoes but not the carrots. She usually gets several successions of fast-growing crops out of whatever starts the season as a bed of carrots. That takes a little surface digging between each crop. The tomatoes, once planted, are there for the season.

Garden Woman decides she will focus primarily on the carrots since they have more stringent requirements. She will plant her carrots in finely prepared soil in beds in early spring, as usual. Then she will plant tomato transplants in a single row down the middle of the carrot beds later in spring after danger of frost is over. That way she can save all the space a tomato patch normally takes, at least early in the season. The tomatoes still won't be "free" because once they are established she won't be able to plant additional successions of fast-maturing crops. The tomato patch will be

pure tomatoes the last half the season and will take up the usual amount of space. But she should be able to save space early in the season. She makes sure the carrot-tomato patch is somewhere she did not have tomatoes last year. So far so good.

In late May when it is time to plant the tomato transplants, Garden Woman finds that only a few of her carrots are ready to harvest. Mostly, she is digging up growing carrot plants in order to make planting holes for the tomato transplants. She trashes quite a lot of carrots, those representing the planting hole as well as where she puts the soil temporarily while digging each hole. Worse yet, it turns out that you really cannot transplant something into the middle of a bed some distance from where your body is all that comfortably or efficiently. Even with far more stretching and reaching and bending and physical discomfort than transplanting usually involves, Garden Woman trashes about a quarter of the carrot plants, far more than just those she removed from the tomato transplants' planting holes.

All then goes well for a while. The tomato plants get big enough to put up the fence-ring cages Garden Woman usually uses for support. Uh-oh. There is no way to get the fencing in there without trashing more carrots. There go another quarter of the carrots.

Meanwhile, it has become obvious that some of the carrots are definitely lacking in the tomato-loving department. The carrots that are closest to the tomatoes are showing little love or even life. They are thoroughly stunted. What carrots seem to love most about tomatoes is being as far from them as they can get without leaving the patch entirely.

In due course, however, Garden Woman harvests the surviving carrots. The carrot-tomato patch reverts to tomato patch. Garden Woman breathes a sigh of relief, mulches her tomatoes, and reverts to watering just once a week.

Later in the season, Garden Woman finds herself running out of ground to plant radishes, scallions, greens, and other things that require a fine seedbed because she used up that prime space on the tomatoes that didn't need it. Meanwhile, the section the tiller skipped that could have supported tomatoes is empty and is just growing weeds.

Mid-August. There is no getting around it. The tomatoes just aren't ripening. The vines are full of big green tomatoes that are just sitting there. Apparently the fertilizer regimen for the carrot patch is too nitrogen-rich for the tomatoes, at least this year. There are no ripe tomatoes until late September. By then the fall rains and cool weather start, and none of the tomatoes are really prime.

So goes the first year of Garden Woman's companion planting adventure. Garden Woman spends lots more work to get lots fewer carrots and tomatoes, gets no tomatoes most of the tomato season and no really prime tomatoes, and has less garden space for crops that need a fine seedbed. But Garden Woman really loves the vision of everything loving everything. She is determined to give peace one more chance. So she tries companion planting again the next year. This time she decides to focus on the tomato patch and treat it as much as possible like a pure tomato patch. She will compromise. She will grow the early planting of carrots in a monoculture. But she will intercrop tomatoes and a later succession of carrots. She will plant the tomato transplants first, then broadcast carrot seed between the tomato plants. How will carrots do treated that way? Can she get some "free" carrots if she treats the tomato patch just like a tomato patch?

Garden Woman plants the tomatoes in a section where the soil didn't till up well and is lumpy. She doesn't fertilize at all. After she plants the tomato transplants, she sows carrot seed very heavily between and around the tomato plants. She knows most of the carrot seed isn't going to come up because the ground is so cloddy. She sows the carrot seed in the areas somewhat away from the tomato transplants and between them, the areas that the tomato plants won't be using for a while. With a little luck, the carrot crop will be ready to harvest before the tomato plants invade the carrots' space and go after all their lunches.

Garden Woman usually just waters the planting positions of her tomato plants once or twice in the week or so after transplanting, then forgets about watering for a month. The deep-planted transplants don't care that the surface of the ground dries up. Their roots are down where the soil stays moist early in the season. If the surface of the ground dries up, so much the better, because that prevents most weed seeds from growing. The tomato patch needs hardly any weeding early in the season, and the tomatoes shade out most weeds later in the season. Not watering the tomato patch early on except right at the planting holes makes for a pretty labor-free situation. But carrot seed will only come up if it is very near the surface in the area that dries out if not watered almost daily. In addition, carrot seed takes a long time to germinate, usually three weeks. If Garden Woman sticks to her usual tomato-patch-watering regimen, *none* of the carrot seed is going to germinate. So she compromises and tries to keep the entire surface of the soil in the patch moist. By a week or so in, she has lost track of this new aberrant need to water the tomato patch daily and isn't. Meanwhile, the regular surface watering, while not enough to germinate the carrots very well, certainly has germinated and started a very solid patch of weeds. Carrot seedlings don't compete at all well with weeds. A month after planting Garden Woman needs to either hoe the patch, sacrificing whatever carrots there are, or to hand-weed. Running a hoe through the entire patch will take perhaps fifteen minutes standing up straight and comfortably. Hand weeding the entire tomato patch carefully enough to notice and salvage any little bitty carrot seedlings submerged in the weeds will take hours on hands and knees. Garden Woman seriously searches

the patch and finds only seven tiny weed-overgrown carrot seedlings, though there are probably dozens, even hundreds more. Maybe. With so many weeds of all kinds it's a bit hard to tell exactly what is a carrot. "*%$#!," says Garden Woman as she reaches for her hoe. "Those seven are probably just Queen Anne's lace anyway."

Garden Women never does find out whether the carrots that were in the tomato patch but quite far from the tomatoes would have loved tomatoes if they had lived long enough. What she *does* learn is that if she has to grow carrots and tomatoes together in one patch, she hates them both.

Polycultures

The best crops for intercropping are those pairs of crops that have similar needs with respect to soil preparation, fertility, crop rotation schemes, watering, and weeding. Ideally the crops also have special agronomical complementarities. One is shallow-rooted, the other deep so the root systems compete less. One is tall and needs full sun. The other grows low and likes shade. One is erect and sturdy and can provide support. The other is a vine that needs support. Whatever weeding regimen works for one also suits the other. Ideally, the pair of crops save you labor when grown together, or at least doesn't cost you too much extra labor compared with the two crops when grown in monoculture. The classic combination of corn and pole beans, for example, is wonderful because it saves labor. You don't have to put up poles for the beans.

In evaluating the yields obtainable with polyculture versus monoculture you have to compare the two crops grown at optimal densities in monoculture with the polyculture at optimal density, not just choose some arbitrary density for everything. There are lots of reports in the agroecology literature testing intercropping by people who would clearly prefer that it be proved advantageous, that is, capable of yielding more than the crops when grown in monoculture. R. Ford Denison provides an excellent summary of this literature and his own thorough analysis in his book *Darwinian Agriculture: How Understanding Evolution Can Improve Agriculture*. In general, in the experiments showing the most advantage for polyculture, the best intercrop combinations yielded somewhat better than the lower-yielding of the two crops in the combination when grown in monoculture, but not as well as the higher-yielding crop in the pair. And even in these cases, the comparisons that made intercropping look best featured spacing so wide for the monocultures that there was bare ground between the plants. They clearly were not at optimal density, causing the advantage of the polyculture to be overestimated and the results invalid. This analysis corroborates my more informal experience. I've concluded that when operating on a bigger garden or field scale, there is little or no yield advantage in interplanting, even optimally, and that it requires planting at many densities to get one planting that is the optimal density for the polyculture in that particular piece of ground that particular season. But even more important, there is usually a huge cost in labor. I think there are sometimes reasons to interplant in the larger garden or farm field, but not usually because of any advantage in yield.

We almost inevitably intercrop in the smaller garden, however, and no fancy quantitative analysis is required to see why it is advantageous. If we harvest part of a bed of something and replant it immediately (with something else) instead of waiting until the entire bed has been harvested a month later and can be planted with another monoculture, we have gotten several square feet of extra growing space for that month that we would not have had if we stuck to nothing but monocultures. It all adds up. Another factor is that in the small garden, where we are usually tending the plants by hand anyway, the polyculture frequently adds little or no additional labor.

Some crops are delightful polyculture combinations whether they give better yields of both crops or not. When I interplant corn and pole beans, for example, I give the corn just a little more space than when I grow it in monoculture. I plant the corn rows at 3.5 feet (1.1 m) apart without beans but 4.5 feet (1.4 m) with beans. In addition, I leave or create occasional gaps in the corn rows when planting or thinning to let more light into the patch. The fewer corn plants give somewhat more and somewhat bigger ears per plant, so the corn yield is about the same even though the plant density is lower. The pole beans yield fewer beans than if they were planted alone and trellised, but I don't care. They yield well enough to be worth my while planting and tending them, and they don't actually cost me any space. The polyculture doesn't cost me any corn yield, and the beans are free. But best of all, I get the greater productivity and bigger size and wonderful flavors of the pole beans compared with bush beans without the labor of putting up poles or trellises.

In my experience, growing squash in corn doesn't work. In my region squash needs full sun. Shaded squash grows much more slowly. If vigorous vines do establish themselves, they also tend to knock over the corn plants.

Cucumbers can be grown on the corn plants at the edge of the corn patch where they can climb up the corn but still get enough sun. Tall-growing nasturtiums don't actually climb, but they can be supported by an edge row of corn and add bright flashes of color as well as culinary diversity. The peppery leaves are great in sandwiches and salads, and the flowers are edible. Small corn-row-wide blocks of peas can grow wherever there is a gap in the corn, and later plantings seem to actually appreciate the shade cast by the corn. The peas bind together into a column that is supported by the corn. If you like the wild and unruly look (which I do), you can interplant a corn patch so as to end up with a delicious, nutritious, beautiful food jungle. Don't forget to leave a space in the middle of the patch unplanted so you can just lie down in the shade of your food jungle and gaze happily at the sky. Or take a nap. Or sit in the shade with visiting friends eating sweet corn raw off the stalks, accompanying it with slabs of cheese and a bottle of wine.

Greens of many sorts can be grown in between rows of corn, as they generally tolerate partial shade. I sometimes grow beds of eat-all greens between alternate rows in the corn patch once the corn is up and has been weeded a couple of times. The eat-alls have the advantage that they grow fast enough and are harvested quickly enough so that they do not require weeding. Anything that requires weeding is hard to tend when planted between the corn rows, even when planted just in alternate rows.

In tropical areas potatoes or sweet potatoes are sometimes planted between corn rows. I suspect this combination requires more direct sun than we have in temperate climes.

Also in tropical areas, it's common to grow taller trees as an overstory with fruit trees as middle or understory. However, many tropical fruit trees are understory trees in nature. Temperate fruit trees of our prime big-fruited types all do best with full sun. They aren't natural understory trees. I think we don't have direct or strong enough sun to raise tree fruits as an understory polyculture without a considerable cost in quality of the fruit. In fact, much pruning is designed to provide extra sun to the fruit even when the trees are growing in full sun. And often it is the fruit that is growing in full sun on the south side of the tree that is tastiest. I think our fruit trees in temperate climes are sun-limited, and need to be the top story in any polycultures, not a mid- or understory. (I don't know whether these thoughts apply to fruit trees in the more southerly states in United States, however.)

There are usually downsides to intercropping. With the corn-and-pole-bean combination, for example, I must wait until the corn is high enough before planting the beans, and end up having to plant the beans later than I would prefer. (If the two crops are planted simultaneously the beans overgrow and shade out the corn. So I give the corn about a month's head start.) This means I either lose the early part of the pole bean season or must plant an early bush bean to fill the gap until the pole beans start bearing.

Interplanting always interferes with optimal crop rotation practices. For most crops, I don't worry about this too much. However, with members of the Nightshade family (Solanaceae), rotation is essential. This reason alone is usually enough to dictate that the nightshades be grown in patches by themselves or with other nightshades. I cheerfully mix tomatoes, peppers, and tomatillos together, however, often using pepper or tomatillo plants as markers between blocks of different tomato varieties.

Supporting and Nurturing

If you are growing determinate tomato varieties you may be able to eschew tomato cages or other supports and just let your tomato plants grow however they want and sprawl on the ground. I prefer to give even the bush types some support. With unsupported tomatoes, between the natural growing patterns and the weight of the fruit, many tomatoes end up resting on or touching the ground. In my garden slugs, sow bugs, rodents, rabbits, birds, and other pests usually take bites out of these low-hanging fruits long before they are fully ripe. The damaged spots then rot or mold. I can get lots more prime tomatoes from fewer plants if I support the plants so that the fruits are all away from the ground. In addition, late blight and fungal diseases spread less readily when the vines are up off the ground and the foliage dries out faster after rains or watering.

When I plant my transplants I leave them unsupported initially. It's easier to tend and weed without the supports. I put the supports in only when the plants are big enough to need them. For determinate varieties I usually just use the modest supports provided by wire-ring-style commercial tomato supports. For the indeterminate varieties I use those commercial supports initially, then add more serious support as the plants need it.

Some people make the more serious cages their indeterminate tomato plants need by making cylinders about 2.5 feet (0.8 m) across with heavy fencing that is 6 feet (1.8 m) tall and has holes big enough to reach through easily. They cage each plant in a cylinder of fencing. I used to do it that way. It works well, and is especially useful in small gardens. Once a week you push all the vines that have started to escape back into their cages. You have to support the support by pounding a T-post solidly into the ground and wiring the fence cylinder to it. Otherwise a serious wind can blow the whole structure over. These tomato cages support the plants much better than anything you can buy, and are especially ideal where you have limited space.

These days I have a bit more space, and I find the constructing of fence-cylinder tomato cages and the staking necessary to put one around each plant a bit too laborious. Instead, I use a combination of the commercial ring supports and stiff wire cattle panels. I plant the indeterminate tomato plants in a row and leave them unsupported initially so they are easier to tend and weed. As soon as the plants are big enough to actually need some support, I poke commercial ring-style supports, the larger size, into the ground around each plant. When the plants overrun those supports, I put up a row of cattle panels behind the entire row to provide additional support. The cattle panels are 16 feet (4.9 m) long and 4.5 feet (1.4 m) high. I use two metal T-poles to support each panel. I used to use two cattle panels, one on each side of the row, to give more of a cage effect, but just one turns out to work well enough when combined with the ring supports. I put small bamboo stakes here and there between the rings as needed where some nice vine is a little short of support. The tomato vines that are heading sideways or back toward the cattle panel I encourage, occasionally moving one here or there so that the panel, ring supports, and other plants all support one another. I let the vines grow through the cattle panel then, when they grow long enough, loop them back through. Occasional vines that head straight away from the row in the wrong direction I just cut off. The vines aren't controlled quite so nicely as they would be if I put fence-cylinder cages around each plant, but the system works well enough, involves less labor, and leaves the plants more accessible for tending and harvesting.

I usually space the tomato rows with about 7 feet (2.1 m) or more on each side rather than my usual 3.5 feet (1.1 m) between rows. The wall of foliage ends up being about 4 feet (1.2 m) wide and casts a heavy shade. It's also too dense and high to let water through easily and creates a rain shadow behind it, so it's necessary to arrange the watering so that the row is watered from both sides.

When I first started gardening, I read all about "tying up the tomato vines." This involves using soft twists or cloth to tie each vine of indeterminate tomato plants to a supporting trellis. The process is usually combined with "suckering," that is, removing suckers, the shoots that form at the nodes between leaves and stems that make additional branches when left to grow. You have to keep tying the vines up and suckering as they grow. I actually did this suckering and tying the first year I grew tomatoes. Then I noticed that none of my neighbors were doing all this work. They just caged their plants and poked the escapee vines back in their cages once a week, and they were getting lots more tomatoes than I was that were every bit as good as mine.

Some experienced growers do sucker their indeterminates; others don't. (No one suckers determinates.) It's said that removing suckers gives you earlier or bigger fruits on the indeterminates. However, you could spend your entire life removing the suckers from not very many big indeterminate tomato plants. Or at least that's what it feels like. I do remove the suckers from the transplants and sometimes the first suckers that are too low to the ground to produce airborne rather than ground-supported fruits. That is, I remove enough suckers to get the plant to put its initial efforts into growing upright and into the supports so as to produce clean, off-the-ground fruit. Beyond those initial suckers, I just let the plants grow undisturbed. However, I understand that those who grow tomatoes in greenhouses usually grow indeterminate varieties pruned to two vines by careful and constant attention to tying up and suckering. There are lots of ways to grow tomatoes that work. All any of us has to do is find one of them.

Sometimes tomato plants start flowering while the plants are much too small to support the growth of both the fruit and the plant. Sometimes there are flowers or even small tomatoes on the transplants. I remove those flowers and fruits. Fruits formed from flowers on small plants or transplants are usually off-flavored in addition to slowing the growth of the plant. Tomato plant babies should not have babies.

Watering and Mulching

Tomato plants start off with deep root systems because you put the transplant in a hole. Then the root system grows even deeper and spreads aggressively too, and additional roots grow from every buried node. Tomato plants (especially big viney types) are less sensitive to drought and irregularities in water availability than many garden plants. However, if the plants get seriously dehydrated, it not only slows their growth but makes the fruits dehydrated as well. Then when it next rains or you water, the fruits that are within about three days of being ripe swell and split. Stress of any kind usually also affects fruit flavor. So to get prime fruit, we need to avoid irregularities in the water supply. One approach is to water regularly. The other is to mulch. When my tomato patch was a small row in my backyard, I did both. I watered the tomatoes at least once a week. In addition, I put down a modest one-season layer of mulch. If I didn't mulch I needed to water the tomatoes twice a week.

I leave my tomatoes unmulched early in the season. We have long cool springs. If I put down even a light mulch, the soil stays cooler and the plants don't grow as fast. About late June I run a hoe through the patch if necessary, then put down about a 3-inch (7.6 cm) layer of straw or leaves. The mulch layer is shallow enough so that I can overhead-water through it. (It's hard to water through a deep mulch layer. You mostly just hydrate the mulch and never reach the soil or plant roots.) In addition, I can still use my big peasant hoe on any weeds, though not any other kind of hoe. With a deep mulch layer, no hoe works and weeds must be hand-pulled. Even a light mulch layer prevents the germination of most weed seeds. Not even a deep mulch layer slows perennial weeds very much, since they regrow from big chunks of root.

The shallow mulch helps in three ways. One is that it protects the soil from drying out. The dry air and wind never reach the soil surface. A second effect is that, with a mulch, the surface of the soil stays moist, increasing the effective depth of the layer of soft, moist soil that never dries out—the optimal zone for plant roots. If, for example, my tilling went down 6 inches (15.2 cm) and the top 2 inches (5.1 cm) dries up between rains or waterings, the soft soil it's easiest for plant roots to occupy is only a 4-inch (10.2 cm) deep layer. With a mulch the soft layer available for plant roots is 6 inches (15.2 cm). Finally, during the growing season the mulch decomposes into the soil, improving and softening it deeper than the layer softened by tilling. Earthworms eat the mulch and poop it throughout their tunnels that go deep into the soil. By midseason the layer of soft soil is deeper than what I started with. By winter there is just a slight skim of mulch left, the rest having vanished into and contributed its virtues to the soil.

Now that my garden is a bigger field away from home, I tend to not have materials around to mulch with, and I do have more trouble with tomatoes splitting. However, the problem is ameliorated by picking all ripe and near-ripe tomatoes before every irrigation, which I do to get the best flavor anyway.

Why It Will Soon Be Impossible to Grow Our Current Generation of Heirloom Tomatoes and What to Do About It—Late Blight 101

Late blight is caused by *Phytophthora infestans*, the same species that was the bane of the Irish in the Great Potato Famine. Late blight is capable of killing tomato plants dead in just a couple of weeks and blasting entire fields into blackened remnants that look like the aftermath of a fire. *P. infestans* is a water mold that is adapted to a land niche. We've recently discovered from genome studies that the water molds are not actually fungi, but just resemble fungi superficially. So in the older literature you will see *P. infestans* called a fungus, whereas now you will see it called an oomycete disease. All the devastation of the Great Potato Famine was caused by one strain of one mating type of *P. infestans* reproducing asexually only. That is, the disease that destroyed every potato throughout Ireland and much of Europe did so with one hand tied behind its back and both legs shackled. Now the ropes and shackles have come off, and we will be living in interesting times. During the last two decades multiple and much more virulent lines of *P. infestans* of both mating types have migrated from their place of origin in Mexico to many other regions. We can expect late blight to become an increasing problem as the new lines spread and become endemic throughout the world. In addition, sexual reproduction will allow the disease to evolve much faster and adapt more rapidly to measures we take to deal with it.

Asexual spores of *P. infestans* live only briefly except on living tissue. They live only long enough to blow to the next living tomato plant in the field or in the region. At the end of the season in temperate zones when the tomatoes die, the *P. infestans* dies with them. The disease survives the winter by living in potato tubers. So as long as we were dealing with just one mating type of *P. infestans* and practiced good potato hygiene, blight on our tomatoes was not too serious a problem. We simply culled our potato seed and plants properly and eliminated volunteer potatoes to keep from infecting our tomato patch at the beginning of the next season. And we watched where we bought tomato starts from if we didn't grow our own. This didn't prevent blight spores from blowing in from elsewhere and infecting our plants, but at least our field didn't start off infected. The asexual blight spores don't survive in the soil.

Sexual propagules of *P. infestans*, called oogonia, are long-lived, however. They persist in the soil. When lines with both mating types become more common and the infestation can reproduce both sexually as well as asexually, some land and some fields will be infected with blight long-term. This will make for a disease that will be much worse and more difficult to manage than anything we have seen to date. So far, there is little evidence

of sexual reproduction of *P. infestans* in temperate North America. This will undoubtedly change in years ahead. The variability of strains of *P. infestans* in Europe suggests that the disease is already reproducing sexually there.

In parts of Europe it is already impossible to grow tomatoes outdoors unless the varieties have serious late blight resistance. I predict that will soon be true for most of the rest of the world as well. All our much-beloved heirloom varieties of tomatoes are doomed unless they have serious resistance to late blight. It's quite possible that an occasional heirloom variety will be blight-resistant. But we already know that most heirloom tomato varieties don't have practical levels of resistance toward the modern lines of blight. If we want to save our heirlooms, we are going to need to breed them for late blight resistance. Furthermore, we can't hold our breath waiting for the university-based plant breeders to do it for us. There are very few university-based breeders compared with a generation ago, and their first priority will necessarily be toward breeding commercial-style varieties—varieties designed to save the commercial market and the tomato processing industries. If we gardeners want to continue growing and enjoying tomatoes with the spectacular flavors to which we heirloom tomato growers have become accustomed, we are going to have to breed blight resistance into these types ourselves. Fortunately, both the major genes associated with late blight resistance are dominant—that is, they express themselves in one dose in crosses. This means those genes will be relatively easy to transfer into our heirloom varieties using traditional plant breeding methods, as I describe and encourage all gardeners and farmers to start doing in the final chapter.

Dealing with Late Blight

Late blight in tomatoes or potatoes spreads by wind or water or physical transfer such as on shoes, hands, or tools. The disease spreads most rapidly under cool, wet conditions. Optimal temperature for its spread is 64 to 72°F/18 to 22°C. Below 59°F/15°C growth of the late blight disease is impaired dramatically. Mildly infected plants produce tomatoes with brown blemishes that make the fruit unpalatable. However, the plants are generally killed pretty rapidly, sometimes in just a few days.

Commercial tomato growers use antifungal sprays as part of their blight-fighting tactics. Organic growers technically can use copper antifungals against late blight, but these poison many microbes and are environmentally damaging. Whether acceptable to formal organic standards or not, any kind of generic poisoning of the microbial life in the garden is counter to the principles of deep organics. Here are some tactics for organic gardeners and farmers.

Either grow your own tomato transplants from seed or buy transplants raised locally by farmers or nurseries you trust. Do not buy transplants from big-box stores. These stores may be importing their transplants from southern areas where late blight can survive the winter on living plants and weeds and infect the transplants. Some of the major blight episodes in the Northeast in recent years are suspected to have been started by imported southern-grown tomato transplants. If you buy transplants, inspect the stems and leaves of each for the scars and lesions associated with blight before you buy. Inspect the vendor's entire plant display. If some of the plants look diseased, even if they aren't the ones you were thinking of buying, don't buy. If the vendor has late blight in some tomato plants, there is a good chance that all are infected, even those that currently appear healthy. Don't worry about whether some funny-looking spots are late blight. You really don't want to buy or import *any* diseases, late blight or not. Buy only blatantly healthy transplants with no lesions of any sort on their leaves or stems.

Late blight is not spread by tomato seeds, not even seeds from plants infected with late blight.

If you buy supermarket tomatoes, don't dispose of them or their debris in your compost. Supermarket tomatoes may come from Mexico, and might carry new strains of late blight from the disease's homeland.

Learn to recognize the lesions associated with late blight as well as how to inspect for it in the field. Late blight normally appears first on the top of the plant, as lesions on the leaves that cross the center vein, and as blotches on stems. There is usually a lighter ring around each brown or black lesion. Cull any plants showing those lesions. Use disposable gloves when culling plants and put the plants somewhere they cannot release spores. Don't compost them. Clean any tools you use in culling. Late in the season of a blighty year when every tomato plant in your region is affected culling isn't practical or useful.

There are good photos of late-blight-infected tomatoes at Vegetable MD Online (vegetablemdonline.ppath.cornell.edu), sponsored by the Department of Plant Pathology at Cornell

University. An excellent video for showing you how to inspect and scout for late blight in both tomatoes and potatoes in the field appears at USAblight.org ("Identifying and Scouting for Late Blight on Farms"). The USAblight.org website is also where to go to report late blight as well as to receive reports as to the existence and spread of blight in the current season in your area.

Avoid spreading blight from your own potatoes. One major way late blight can survive the winter is in infected potato tubers. *Cull all volunteer potato plants.*

If you buy all your seed potatoes, buy only certified seed. However, certified seed is not guaranteed disease-free. You should still inspect the tubers carefully and watch the potato patch and cull any plants that look diseased. On potato tubers late blight shows up as sunken black lesions. When you cut across the lesion there is a dry, firm reddish-brown rot. However, late blight lesions often allow co-infection with other diseases, so their appearance is not always typical. Just cull all potato seed with any signs of rot or disease of any sort or whose sprouts look at all different from what is expected (another sign of disease). Lines of blight that are very tomato-adapted sometimes may not show any obvious signs in potato tubers, even though they are present. So watch the potato patch and cull all diseased potato plants promptly.

If you save your own potato seed, know what you are doing. In *The Resilient Gardener* there is a large chapter, a mini book basically, on potatoes. It has a long section on saving your own potato seed that teaches you to use the same kinds of screening and culling methods used by certified potato seed growers, only with the methods made even more powerful by my incorporating additional tricks practical only on the small scale. The days in which it is acceptable to blindly save your own potato seed without knowing how to recognize and screen against diseases in both growing plants and tubers are over. Whatever the source of your potato seed, cull any diseased potato plants.

There will be increasing pressure on gardeners and farmers to buy all their potato seed and not save potato seed at all. If we have to buy all our potato seed that will be the end of growing heirloom potatoes. By and large, the starting seed is not available at prices that are affordable as other than small starts to multiply into enough to grow as food in future years. So those university and seed company guys who are so blithely telling us to never save our own potato seed are actually saying: "Give up heirloom potatoes. Let them vanish into the dust. Grow only the few (pallid-tasting) commercial varieties we see fit to provide as commercial certified seed. And eschew independence. You must buy new seed potatoes from us every year. And by the way, we don't care about all you small market farmers who grow and sell heirloom potatoes. You aren't real agriculture to us. We really don't care if we deprive you of growing what makes you unique and destroy your ability to survive economically." I suspect we should be watching out for laws that forbid saving our own potato seed.

I wouldn't bother growing potatoes at all if I had to grow just commercial varieties. In addition, growing lots of varieties, especially varieties that are different from the main commercial varieties in our region, is an important part of our strategy for dealing with the threat of potato diseases of all kinds, including late blight. Usually some of the varieties will have some resistance to the epidemic du jour. If the nineteenth-century Irish had been growing dozens or hundreds of varieties instead of just one variety that turned out to be unusually susceptible to a particular strain of late blight, it's likely that the Irish Potato Famine would never have happened at all, or at the very least would not have been so devastating. Andean farmers typically grow dozens or even hundreds of different varieties of potatoes in each potato field. So should we.

Buying all our potato seed would also eliminate potatoes as a practical survival crop for the family or homestead. It would mean that if the times get rough enough—if we are out of money, for example, or there is a breakdown of any sort that cuts us off from commercial certified potato seed—we would lose our ability to grow potatoes just when we needed the crop the most.

The same university people who are telling us not to save our own potato seed go to third-world countries to teach them how to save theirs. I encourage you to save your own potato seed, but I exhort and entreat you to learn to do it right.

I don't water my potato patch, which not only gives me much more flavorful potatoes, it lessens the problems with fungal and oomycete diseases as well as pests of all sorts, giving much cleaner potatoes. In addition, my potatoes, being non-irrigated, are actually never very close to the tomatoes, which are in the irrigated garden. That isolation would probably also help if I got late blight in my potatoes. If you grow

potatoes and live anywhere other than the arid Southwest, I suggest experimenting with growing your potatoes without irrigation.

While some lines of late blight can spread between potatoes and tomatoes, other lines seem to have strong preferences for one or the other. In some cases, both the potatoes and tomatoes in a field have late blight, but have completely different strains.

Avoid tromping through a blight-infected garden and carrying blight back to your own. When I visit Oregon State University's blighty blight-resistance selection fields, I wear shoes I don't wear to my own garden. And I change clothes and shower when I get home before going to my own garden.

Hoe, tend, and harvest tomato plants when they are dry, not wet. Both oomycete and fungal diseases spread most readily when the plants are wet.

Grow lots of different varieties. Diseases usually spread fastest in monocultures. If you have many varieties, the disease that has just adapted to grow in one variety optimally is not adapted to grow optimally in the next to which it spreads. In addition, you are more likely to be growing some variety that is resistant to the line of blight that is in your region that year. In the future, as preparation for times coming, I'm going to start growing a much larger repertoire of tomatoes. Next year I think I'll plant a dozen varieties with just one plant of each in addition to my regulars. I'll plant the new varieties in rows without support and just give the support to my regulars. That way the new couple of dozen will cost me minimal work. (I won't care whether I get as many prime tomatoes as possible from each plant, since I don't even know whether I'll like the tomatoes compared with my regulars and will not really be counting on them.) If some variety tastes as good as my regulars, or does particularly well—for example, staving off late blight when others succumb—it will earn its way into the supported section next year. Meanwhile, the biodiversity will give me some insurance against diseases of all sorts for minimal effort, and I'll be able to accumulate information about the blight resistance of some of the heirloom varieties.

Grow a bigger proportion of early varieties. In a bad late blight year, the early varieties may give you a crop before late blight takes over.

Start experimenting with blight-resistant varieties, both modern varieties bred for blight resistance as well as heirlooms documented as having some resistance or even just rumored to have resistance. The most blight-resistant modern varieties are, alas, hybrids. If you like their flavor, start dehybridizing them. If you don't like them as well as your favorites, cross the resistant hybrids to your favorites and start creating your own resistant varieties, as I describe in the last chapter.

The new practice of grafting tomatoes onto "disease-resistant" rootstocks will not help with late blight. Those rootstocks are not resistant to late blight. Even if they were, it wouldn't help, because late blight infects foliage and stems, not just roots. You would still need for the top of the plant to be blight-resistant.

The "disease-resistant" rootstocks are resistant to soilborne diseases, the diseases we home gardeners can avoid just by rotating our crops.

Finally, use water and moisture management to prevent or slow the spread of late blight. Late blight is a water mold. The vegetative zoospores that spread the disease actually swim for a bit on the surface of the tomato leaf before invading it. They need the leaf to be wet or at least thoroughly moist for about eight to twelve hours in order to cause an infection. The moisture can come from rain, watering, dew, or even fog. The pathogen's very nature as a water mold is its greatest vulnerability. So we can use water and moisture management as part of our strategy, especially in areas where irrigation, not summer rain, is the source of water.

My first line of defense against late blight is the watering pattern I've adopted to help control powdery mildew in the squash patch. I no longer water my irrigated garden at night or in early morning. Watering at night or in early morning saves water, but it means the plants and ground are wet or moist for many hours, making perfect conditions for the spread of late blight as well as powdery mildew and other fungal diseases. Instead I water during the day or early evening when the plants will have some time to dry out before the dew comes on in the early-morning hours. With the old watering pattern my squash patch would usually die some time in midfall from powdery mildew. With the new watering pattern, my squash patch is much less affected by late-season powdery mildew. It appears weeks later and affects the plants less. The plants keep producing and don't actually die until they get blasted by cold weather. I'm sure the daytime watering pattern will also slow the spread of early blight, late blight, and many other diseases.

Presumably, if you manage your tomatoes so as to have to do less irrigating, that would also slow the spread of late blight. Thus if you mulch your tomatoes, and this means you water them less frequently or not at all, this might be a blight mitigator. You would not have to wet the foliage as frequently. Drip irrigation instead of overhead watering might also help by keeping the foliage dry.

Anything you do to keep the plants from being moist longer than necessary is a help. So supporting plants so that the foliage dries out quickly after rain or watering is helpful, as is not planting the plants too densely, thus ensuring good air circulation.

If your tomato planting is in a moister area, such as a low-lying area or a spot shaded by a nearby woods, those plants will be more vulnerable to late blight. It's best to put the tomato patch in a well-drained area with full sun.

Late Blight Resistant Hybrid Tomato Varieties

University and seed company breeders in The United States and elsewhere are working hard to breed late blight resistant tomatoes. But alas, even public-funds-supported university breeders these days often present new releases only as hybrids or proprietary (patented or PVP) varieties. In addition, these hybrids are commercial tomatoes, and are likely to include genes for slow ripening, tough skins, uniform color, and other characteristics that make for a visually appealing tomato with a long shelf life but inferior flavor. If you like these varieties, you can dehybridize them. If their flavors don't measure up to what you want, you can use these varieties as a source of genes to cross into varieties you like better. *Ph2* and *Ph3* are the two genes for blight resistance that have been genetically characterized. (*Ph1*, the first late blight resistance gene to be discovered and studied, is not effective against modern lines of late blight.) *Ph2* and *Ph3* are unlinked dominant genes. In the list below I indicate which genes are present in each hybrid and whether they are present in one or two copies, that is, whether they are heterozygous or homozygous, where that information is available. We will make use of that information in the final chapter when we discuss how to dehybridize these varieties or transfer their genes for blight resistance into our favorite varieties.

I also indicate resistance to early blight and other diseases by the code at the end of each listing. (AB stands for alternaria blight, that is, early blight. LB is late blight. F, F2, and F3 indicate resistance to race 1, races 1 and 2, or races 1, 2, and 3 of *Fusarium* wilt, respectively. V is for resistance to *Verticillium* wilt. SLS is septoria leaf spot resistance. N is for nematode resistance. TMV is tobacco mosaic virus resistance. TSWV is tomato spotted wilt virus resistance.)

'Hybrid Jasper'. 60 days. IND Red cherry. Up to 0.35 ounce/10 g. Fruits in trusses. Late blight resistant; genetic basis unknown. Intermediate resistance to early blight. AB F2 LB.

'Hybrid Golden Sweet'. 60 days. IND Yellow grape. Up to 0.7 ounce/20 g. Fruits grow in long clusters. F LB.

'Hybrid Juliet'. 60 days. IND Red cherry. Mini Roma. Up to 2 ounces/56 g; up to 2.3 inches/5.8 cm. Intermediate blight resistance. AB LB.

'Hybrid Mountain Magic'. 66 days. IND Red cherry. 2 ounces/56 g. Late blight resistant; heterozygous for both *Ph2* and *Ph3*. AB F3 LB V.

'Hybrid Defiant PhR'. 70 days. DET Red. 8 ounces/227 g. Late blight resistant; heterozygous for both *Ph2* and *Ph3*. High resistance to late blight; intermediate resistance to early blight. AB F2 LB V.

'Hybrid Plum Regal'. 75 days. DET Red. 4 ounces/113 g. Late blight resistant; homozygous for *Ph3*. AB F2 LB TSWV V.

'Hybrid Iron Lady'. 75 days. DET Red. Up to 4 inches/10.2 cm. Late blight resistant; homozygous for *Ph2* and *Ph3*. Plants 2.5–3 feet (76–91 cm) tall. Robust resistance to septoria leaf spot. AB, LB F2 V SLS.

'Hybrid Mountain Merit'. 75 days. DET Red. Up to 10 ounces/283 g. Late blight resistant; heterozygous for both *Ph2* and *Ph3*. F3 LB N TSWV V.

'Hybrid Ferline'. 95 days. IND Red. 5 ounces /142 g. European late blight resistant hybrid. Available from Totally Tomatoes and Thompson and Morgan in U.S. V F N T.

'Hybrid Fantasio'. Maturity time unavailable. IND Red. 8 ounces/227 g. European hybrid late blight resistant tomato. Available from Suttons and other UK seed companies; apparently unavailable in U.S. LB TMV V F N.

'Hybrid Iron Lady' is homozygous for both *Ph2* and *Ph3*, that is, it has two copies of the resistance-inducing version of both genes. This will make it especially useful for us in our own breeding work to develop open-pollinated varieties. 'Iron Lady' is available only from High Mowing Organic Seeds.

'Ferline' and 'Fantasio' are European hybrid varieties. I was unable to find any information as to what genes they contain. Only 'Ferline' seems to be available in the United States.

The rest of the hybrids are available from Johnny's Selected Seeds. 'Defiant' and some of the others are available only from Johnny's (in the United States). A few of the hybrids, such as 'Mountain Magic', are being listed by many companies. Veseys in Canada seems to be the sole Canadian distributor for 'Defiant' and also carries some of the other resistant hybrids.

Late Blight Resistant Heirloom and Open-Pollinated Varieties

It is already clear that most heirloom tomato varieties are susceptible to the new lines of late blight. However, there are reliable reports of some heirlooms having significant blight resistance. There are doubtless others that have never been tested. A variety need not be totally resistant to be useful. If a variety has enough resistance to slow the disease by a month, it might give you a full crop or nearly full crop of tomatoes before being overly affected. The genetic basis for the resistance in the heirlooms generally has not been studied, so it is unknown. In many cases where the variety is blight-resistant, it will be only partially resistant or resistant to only certain lines. Because different strains of blight predominate in different years, this means that a variety that is genuinely blight-resistant one year can be blight-sensitive some other year, and vice versa. By growing a number of different varieties we are much more likely to have at least some that are productive in any given year. (Contrary to the assumptions of many fans of heirlooms, most tomato disease resistance genes were not discovered in heirlooms. Most resistance genes were found in wild species and then transferred into culinary varieties.)

Most of this list of blight-resistant or -tolerant heirloom and open-pollinated varieties comes from a compilation in a 2010 Internet article by T. A. Zitter, Department of Plant Pathology at Cornell University, "Keeping Late Blight in Your Rear View Mirror." Some other information is from certain seed catalogs, as indicated. That this list contains mostly familiar varieties undoubtedly reflects the fact that these varieties are the ones that get chosen for testing or happen to be present when a blight epidemic hits a relevant field. We need to grow lots of heirlooms and rarer varieties to expand our knowledge as to which ones have something special to contribute in this new and blightier era. The list is arranged by maturity time from earliest to latest as best I can estimate given maturity times from different regions.

'**Red Pearl**'. 58 days. IND Red grape tomato. Up to 0.7 ounce/20 g. Intermediate resistance to late blight. F2 LB (Johnny's Selected Seeds).

'**Stupice**'. 60 days. Small-vined IND Red. Up to 2.5 inches/6.4 cm.

'**Slava**'. 60 days. IND Red. 2 inches/5.1 cm (Adaptive Seeds).

'**Matt's Wild Cherry**'. 60 days. IND Red. Up to ⅝ inch/1.6 cm. Resistant to late blight; genetic basis unknown. Also some resistance to early blight.

'**Yellow Currant**' (*S. pimpinellifolium*). 60 days. IND Yellow. 1 ounce/28 g.

'**Geranium Kiss**'. Early DET Red. Up to 2 inches/5.1 cm. Dark-green foliage looks like a geranium. Late blight resistant; genetic basis unknown. Resistance gene(s) probably derived from the wild species *S. humboldtii*. Bred by Alan Kapuler and Kusra Kapuler (Peace Seeds and Peace Seedlings).

'**Legend**'. 68 days. DET Red. Parthenocarpic. Up to 5 inches/12.7 cm. Late blight resistant; homozygous for *Ph2*. Bred by Jim Baggett/OSU (Nichols Garden Nursery).

'**Pruden's Purple**'. 67–72 days. IND Pink flavor class; they look red. Many over 1 pound/453 g.

'**Quadro**'. 70 days. IND Red. Plum-shaped multipurpose Roma. Bred by Hartmut Spiess in Germany for late blight resistance; genetic basis unknown (Adaptive Seeds).

'**Black Plum**'. 75 days. IND Black. 2 inches/5.1 cm.

'**Red Currant**'. 75 days. IND Red. ⅜ inch/1 cm.

'**Tigerella**' (aka 'Mr. Stripey'). 75 days. IND Bicolor. Up to 6 ounces/170 g.

'**Old Brooks**'. 75 days. IND Red. Medium-large. Resistant to early and late blight (Totally Tomatoes).

'**Black Krim**'. 80 days. IND Black. Up to 18 ounces/510 g.

'**Brandywine**'. 82 days. IND Pink. Up to 1.5 pounds/680 g.

'**West Virginia 63**' (aka 'Centennial'). 70 days at SESE/Virginia. IND Red. 3 inches/7.6 cm. Bred by Mannon Gallegly. Late blight resistant; homozyous for *Ph2*. Also resistant to *Verticillium* and *Fusarium*.

'**Aunt Ruby's German Green**'. 80–85 days. IND Green. Up to 18 ounces/510 g.

'**Aunt Ginny's Purple**'. 85 days. IND Pink. 16 ounces/453 g.

'**Big Rainbow**'. 90 days. IND Bicolor. 16 ounces/453 g.

Why the Best-Flavored Tomato May Not Be the One That Is Picked Vine-Ripe

In the romantic image the homegrown tomato is always picked vine-ripe. I had not been growing tomatoes for very long before I noticed that in many cases my tomatoes tasted better when they were picked two to four days short of perfect ripeness and allowed to finish ripening indoors. Sometimes the vine-ripe tomatoes actually tasted pretty bad, in fact, and even had distinctively bitter skins. But sometimes not. It didn't seem to have to do with varieties. After serious observation I came to a simple conclusion. When grown in my region, fully vine-ripe tomatoes picked in the morning before they have had much sun on them don't taste very good. Furthermore, on cooler days when there is no real direct sun, the tomatoes also don't taste very good. The best time to pick them is from midafternoon on, not early in the morning, and it must be on a day with some serious sun. The tomatoes picked vine-ripe early in the morning or on cool days with no sun taste quite similar to tomatoes that have been refrigerated.

You are not supposed to refrigerate tomatoes. It is well known that this destroys the flavor and makes them taste awful. The vegetable storage experts tell us to keep picked tomatoes above 50°F/10°C. In the Willamette Valley the night temperatures drop to near 50°F/10°C for both the early and late part of the season. Even during most days in August, the night temperatures are likely to be not much above 60°F/16°C. Is 50°F/10°C a magic cutoff that leaves tomato flavor unaffected above it and damaged horribly below? I doubt it. It's more likely that the flavor begins to be affected at temperatures considerably higher than 50°F/10°C, just not so dramatically.

I think the flavor of tomatoes in the Willamette Valley is harmed by our low night temperatures even in August. When I bring tomatoes indoors, they don't experience these low nighttime temperatures. I do open my windows at night and close them during the day—the traditional Oregon air-conditioning. This cools the place as well as is needed, but it never gets as cold indoors as it is outside at night. If you are in the maritime Northwest or other places with cool nights, I suggest picking your tomatoes in the afternoon after they have had some sun on them, or picking them a bit early and finishing them off where it stays a bit warmer. I do both. The earlier-picked tomatoes are indistinguishable in flavor from those picked vine-ripe after sun. These earlier-picked tomatoes are not anything like supermarket tomatoes, incidentally. I pick them when they have started to turn red, just not all the way to full ripe red. Picked any earlier than that (as supermarket tomatoes are), they will never develop their full flavor.

I've talked with various people in various parts of the country about tomato flavor and picking time—people in places that don't have the cold nighttime temperatures we experience in the maritime Northwest. They don't seem to have this problem with vine-ripe tomato flavor being so much affected by the time of day the tomatoes are picked.

Full-ripe or almost full-ripe tomatoes split when rained upon or watered. Tomatoes earlier than my near-ripe picking stage don't normally split when watered. So I usually pick twice a week, before irrigating, in late afternoon so the full-ripe tomatoes will be prime. I pick all the tomatoes that are full-ripe as well as all that are within about four days of full ripe. I pick into cardboard flats of the sort that grocery stores receive canned goods in. (See the photo section.) Grocery stores recycle large numbers of these flats and are happy to provide all I want. I never stack tomatoes on top of one another. That causes bruising. After I pick, I irrigate. When I get the tomatoes home, I arrange them in the box from less ripe to most ripe to facilitate using the tomatoes at peak ripeness. When I give away flats of tomatoes, I also arrange them that way.

One reason why I consider growing my own tomatoes obligatory is that even farmers market tomatoes of the same varieties as those I grow don't taste all that great compared with those I grow myself and pick optimally. The farmers marketeers make a big deal out of picking them vine-ripe. Because of the logistics, however, they pick them all in the morning. In addition, they pile the tomatoes up. The bruised places are mushy and will deteriorate quickly unless the tomatoes are used almost the same day. In addition, because the tomatoes are fully ripe, they need to be used that day or the next. You only get a day or two's worth of tomatoes from a trip to the market.

When I give away tomatoes to friends I give them away by the cardboard flat, with a lovely monolayer of perfect unbruised tomatoes of maturities suitable for eating from right then to half a week from then. I suggest this is a good way to sell tomatoes in the farmers markets too. That is, pick the tomatoes into cardboard flats and keep them in monolayers. Then sell them by the flat so the customer buys and has perfect tomatoes for half a week, not just tomatoes that are optimal that one day. You would need to rig a method for hauling and displaying the tomatoes in crates, but I think it would be worth it.

If you live outside of the maritime Northwest, the effect of picking time of day on tomato flavor will probably be different from what I experience. Perhaps fully vine-ripe tomatoes will taste best. However, try picking and eating tomatoes at various times. You might have quite different patterns. In very hot, windy, dry areas, for example, the tomatoes might be under a lot of stress in the afternoon. Perhaps those tomatoes would taste better picked earlier in the day. Variety seems also to be a factor. Some varieties seem to be more affected by our cold nights than others. I once went to a tomato tasting of about fifty varieties. They had been picked over a period of several hours starting from about dawn. It was a cool overcast day. I thought none of the tomatoes tasted very good, not even those of varieties I usually like. I think our evaluation ended up having more to do with when the tomatoes were picked than with the varieties.

The way to tell when a tomato is perfectly ripe is partly by color but partly by a little judicious squeezing. Each variety has its own pattern, so you learn how to tell when the tomatoes of a particular variety are ripe. When comparing varieties, you need to compare tomatoes picked at their prime from each variety, not just two arbitrary tomatoes, one picked prime and the other not. Use blind tests and replicates. If the result is real and significant, it should be reproducible.

Most varieties hold well enough to be prime for about three days or so. The paste tomatoes tend to hold better than the other types. I suspect that the extra-long shelf life that commercial growers are so proud of breeding into all their varieties, however, will turn out to be another characteristic invariably associated with inferior flavor.

Many people raise huge amounts of extra tomatoes to put up, that is, can or process into tomato juice or sauce or spaghetti sauce or ketchup. Some people grow large amounts of paste tomatoes for the purpose. Others just use whatever excess tomatoes they have. The only processing of tomatoes I've done is to make frozen tomato juice. For this I dip the tomatoes in boiling water for a few seconds and then into cold water so that the skin peels off easily. I then chop the skinless tomatoes into big chunks, put them in an 8-inch (20.3 cm) Pyrex bowl (without water), cover them, then zap them in the microwave oven for a couple of minutes or until they release their fluid and disintegrate. I freeze the juice in one-meal portions in plastic freezer containers. This juice is just blanched rather than cooked, and has an intense fresh tomato flavor that is destroyed in the longer cooking for canning and most processing. In winter, I add some of the juice to tomato soup or sauces to enliven their flavor.

My regular readers will not be surprised to hear that I tried to make the zapped juice without peeling the tomatoes. It doesn't work. The skins curl into very tough stringy unappetizing strips. Usually people not only remove the skin on tomatoes to make juice or sauces, but also strain out the seeds. I don't mind the seeds, so I skip the straining. 'Amish Paste—Kapuler' is the variety I prefer for making the zapped juice, but all the tomatoes I grow work fine, each making a different-flavored juice, but always wonderful.

I eat nearly all my tomatoes fresh, however. I eat them out of hand. I chop them into salads. I add them to soups or stews right before serving. (They don't need to be skinned since the tomatoes will be warmed rather than actually cooked. The skins don't curl away into unpalatable strips unless the tomatoes are actually cooked.) I just crumble some feta cheese over them and call it a meal. I dress tomato chunks with a little ketchup and balsamic vinegar and Italian seasonings for an all-tomato salad. I put them into spaghetti sauce, so many it is just chunks of warm tomato coated with sauce. Then I put the warm spaghetti-sauce-dressed tomatoes over rice or polenta. 'Black Krim' is my favorite for combining with spaghetti sauce. I suspect 'Black Krim' has more umami flavor components than my other tomatoes. This may be typical of black tomatoes in general.

Once the tomato season is in full swing, I eat large amounts of tomatoes daily as well as give them away. Homegrown tomatoes are my favorite garden crop to use for gifts. By the start of the cooler weather and fall rains, the tomato quality deteriorates. But by then I've generally gotten a bit burned out on tomatoes and am glad to just forget about them until the middle of winter. Then some cold miserable day I remember my frozen tomato juice, and thaw out a container of pure summer.

Using Green Tomatoes

Some people bring the green tomatoes indoors at the end of the season to finish ripening. Some people even grow the special varieties that hold longest when picked green and are best at ripening from a green-stage picking. I don't. To my taste, tomatoes so ripened always have off-flavors to some extent and aren't prime. And by the end of the tomato season, I have usually eaten large amounts of tomatoes nearly every day for weeks and am perfectly happy to stop eating them for a while.

If you do want to ripen some of your end-of-season green tomatoes, you can get the best flavor by bringing whole plants or branches with the tomatoes attached indoors and letting the tomatoes ripen on the vine. If you pick individual tomatoes for ripening indoors, pick those that are full-sized and are starting to lose their chlorophyll. Deep-green tomatoes will not ripen. Store the tomatoes indoors in monolayers with the tomatoes not touching one another. Some people instead wrap the green tomatoes in newspaper to keep them separate enough. The tomatoes continue ripening indoors. The special varieties bred for post-season ripening may take a couple of months to ripen, and keep a couple of months ripe, supposedly. (In my own experience, this post-season ripening gives such poor flavor that I can't really tell what is supposed to count as ripe.)

I've read that some people ripen their green tomatoes as needed by putting them in a paper bag with a ripening banana. The banana produces ethylene gas, the gas that is used to ripen supermarket tomatoes. I haven't tried this. Figuring out how to make supermarket tomatoes at home has not been high on my list of priorities.

Green tomatoes at all stages are edible cooked but not raw. In fried green tomatoes, the classic southern dish, the tomatoes are usually breaded and fried in a serious layer of bacon grease or other fat. There are lots of different recipes for fried green tomatoes on the Internet, some involving flour as well as cornmeal or dipping in an actual batter. I consulted my southern-raised friend Kinsey Green about it. She says they just sliced their tomatoes into slices about a quarter inch (0.64 cm) thick, dredged them in cornmeal with a little salt and pepper in it, and fried them in bacon fat. This probably represents the original style for the dish, because in the American South, corn was the staple grain; wheat didn't grow well there, and bacon fat or pork lard were major staples. The tomatoes are supposed to end up having a crisp, crunchy coating. I've never had them, though I imagine just about anything breaded in cornmeal and fried in serious amounts of bacon grease probably tastes pretty good.

When I use green tomatoes, I just chop them up and use them in soups, stews, or stir-fries the same way I use summer squash. Cooked green tomatoes don't taste anything like ripe tomatoes. Their predominant flavor is just one of sweetness. They are surprisingly sweet, too sweet to use as more than a minor ingredient in most soups, stews, or stir-fries. They could undoubtedly be used to make a good curry, though, where fruit or some other sweet ingredient works well. I would also suggest trying them in sweet-and-sour soups or hot, sweet, and sour soups. I tend not to use green tomatoes because they are available at a time of year when so much I like better is available. However, if you are short on food at the end of the tomato season, just realize that all those green tomatoes are actually food.

Green tomatoes really come into their own for making pickles, relishes, chutney, green tomato ketchup, and other such products at the end of the tomato season. Apparently any green tomato at any size or stage of development will do. The tomatoes don't have to be peeled. In many of the recipes, you add sugar or vinegar, and in some of them you cook or can. But some of the recipes simply lactoferment the tomatoes. Like cooking, lactofermentation transforms the inedible raw green tomatoes into edibility.

When I write a book I always discover some connections I've missed until then. *Why in the world am I not lactofermenting my year-end supply of green tomatoes?* Ordinarily I don't process vegetables because I can't stand leaning over a hot stove in August or September, and I'm too busy then anyway. But lactofermentation doesn't involve any cooking or elaborate processing. And green tomatoes will hold well at even room temperature, and would certainly hold even better in my attached garage. I could pick all the green tomatoes just before last frost and deal with them after last frost when things have settled down and I have more time. I wouldn't even have to pick the green tomatoes into flats as I do with ripe tomatoes, since the green tomatoes would be hard and wouldn't bruise. I could just pick into crates. Then I could just chop the tomatoes into chunks, add some salt or salt brine and garlic and other seasonings, and some tannin-containing leaves like oak or horseradish or grape leaves, and let the fermenting begin. Wow. As I sit here in the middle of winter I find myself looking forward eagerly to . . . of all things . . . next year's green tomatoes.

Stop trouble before it starts. Make order before there is chaos.
Deal with the small before it is large. Deal with the few before they
are many. Begin the difficult while it is easy. Approach the great work
through a series of small tasks. The largest evergreen grows from a
tiny seedling. The journey of a thousand miles starts with a single step.

If you rush into action, you may stumble. If you stand on tiptoe you
do not stand stable. Failure comes most often near completion.
Be as careful at the end as the beginning.

道

CHAPTER EIGHT

Nurturing—Weeding

Avoid, Delay, Remove. Garden Woman Meets Pigweed with Attitude.
The American Square Hoe. Buying, Using, and Sharpening the Peasant Hoe.
Buying, Using, and Sharpening the Coleman Hoe. Stirrup Hoes.
Wheel Hoes. Electric Wheel Hoe and Electric Tiller.

Avoid, Delay, Remove

After providing our plants with adequate nutrients and water the most important thing we must do to properly nurture them is to remove competing plants. Hence we thin so that the plants are not too crowded to grow properly. In addition, we control weeds. Most of the labor of gardening is associated with weed control. You can maintain a pretty good-sized productive garden with just a few hours of work a week if you get good at weed control. There are three basic approaches to weed control—avoidance, delay, and removal.

The most important avoidance method is to not let weeds get away from you. This means not letting weeds go to seed in your garden so you don't have to deal with a much worse situation in subsequent years. It also means pulling or cutting back perennial weeds so as to keep them from becoming better established and spreading.

Another tactic for avoiding weeds is to stir the ground as little as possible when you hoe. If you stir up fresh soil, you bring fresh weed seeds from the soil seed bank into the germination zone. Most small weed seeds have seed dormancy tricks that prevent their germinating when they are buried too deep. The seed might require just the right amount and kind of light for germination,

for example, the light the seed gets when buried beneath a thin layer of soil. If you weed when the weeds are tiny—less than an inch high, for example—you can use light hoes and a hoeing style that barely penetrates the soil. If you wait until the weeds are larger you need a heavier hoeing action that stirs the soil deeper and brings up weed seeds that can then germinate.

A weed-avoiding trick I have used with small garden beds takes advantage of weed-seed-free compost. I spread a 1-inch (2.5 cm) layer of compost on the surface of the bed, both to provide nutrients as well as to control weeds. If the layer is left intact on top of the soil, it is too deep to allow the weed seeds in the soil beneath to germinate. I then sow small garden seed on the surface with just very gentle raking in to bury them adequately without mixing the compost and underlying soil. For bigger seeds like corn, beans, peas, and squash that should be planted deeper, I plant in the soil first, then apply the layer of compost on the planted bed. This trick can give a virtually perfect weed-free bed as long as there are no perennial weed roots in the bed. The method has an added advantage. The black compost on the surface of the bed absorbs more energy from the sun and warms up more than does a bed with a lighter-colored soil surface. This is particularly advantageous in early spring when cold soil limits the growth rate of many plants.

Other weed-avoidance tricks depend on the fact that in land that hasn't been rearranged by digging or tilling for a few years, most of the weeds in the soil seed bank have had time to die. So most of the viable weed seed is on or near the surface of the soil. When I make a garden bed from an area of lawn, I like to start by removing the sod entirely with my peasant hoe. This actually removes nearly all the weed seeds. Since my backyard soil is almost pure clay, I can then work in sand and compost to provide the materials necessary to turn the low-fertility sod-stripped clay sunken bed into a fertile garden loam raised bed. That first season I'm likely to have pretty close to no weeds. If instead I break up the sod and turn it under, I deliver all its viable weed seed to the soil seed bank. That seed may be buried for the first year, but it can come up when the soil is rearranged in subsequent years. In addition, it's not easy to turn sod under by hand digging well enough so that some of it isn't near the surface. Chunks of sod are lots worse weeds than most weeds coming up from seed. If you are creating a bed from lawn using hand tools I really recommend completely removing the sod first and composting it.

Plowing with a moldboard plow is itself a weed-avoidance trick. You actually turn a big layer of soil over completely so the upper part that has most of the weed seeds is turned under too deep to allow the seeds to germinate. Rototilling, which merely stirs up a shallow layer of soil, buries many surface weed seeds, but is not nearly as effective as plowing. In addition, it softens only a very shallow layer of soil, a layer too shallow to be optimal for most garden plants in most situations. Using a moldboard plow is often criticized for its rearrangement of the soil profile and for creating plow pans when used regularly. (Plow pans are compacted layers beneath the surface of the soil where the plow rides as it works.) However, you don't have to plow frequently, nor at the same depth from year to year. And turning the more fertile top layers of soil under doesn't matter too much if the entire tilled depth is fertile or can be made adequately fertile. Chisel plows, which loosen the soil without turning it over, avoid ruining the soil profile and creating plow pans, but also don't have as much weed-mitigating effect as you get when you turn over the soil.

Of course, if you turn over the soil, by whatever means, then let weeds go to seed in the area, then plow or till again each year, pretty soon you have a large, fresh crop of weed seeds that will be viable for years worked throughout your soil so thoroughly that all rearranging of the surface of the soil merely brings up fresh weed seeds.

Another weed-avoidance tactic is using mulch. It takes a heavy mulch—say 8 inches (20.3 cm) or more when compacted—to prevent all weeds including most perennial weeds from coming up from established chunks of root. Vegetable gardening in permanent deep mulches doesn't work for most people in most areas of the country. It keeps the soil too cool to grow plants well in spring and breeds pests of all kinds ranging from slugs and sow bugs to rodents. But even a light mulch—say 1 to 3 inches (2.5 to 7.6 cm)—can prevent most annual weeds that start from small seeds. And you can water through a light mulch, but not through a heavy one. With small garden beds, especially those started from transplants, I often apply a light mulch in late spring after the soil has warmed up and the plants are established. It not only enhances soil organic matter and tilth, but may prevent the need for any further weeding for the rest of the season. A light mulch also increases the effective depth of the bed, since it keeps the surface of the soil from drying out so that plant roots can make full use of it. Even a light mulch interferes with our ability to use most styles of light hoes, however, so it can turn whatever weeding is still needed from an easy hoeing proposition to much more laborious hand weeding. I prefer light mulches about 3 inches (7.6 cm) deep because they can usually stop annual weeds from germinating, and are light enough so that the heavy peasant hoe (though no other hoe) can still weed through them. By the end of the season, most of the light mulch has been incorporated into the soil by earthworms and other critters.

However you start a garden, maintaining it relatively weed-free requires keeping weeds from going to seed in it. However, life happens, and sometimes our gardens get away from us. So most of us end up with at least some sections of our gardens that have their full share of weed seeds in the soil seed bank or that are infested with perennial weeds. So we must also control weeds by using delay and removal.

By "delaying weeds," I mean giving the garden plants a head start over the weeds. The most extreme delaying tactic is to use transplants. We grow the transplants indoors under conditions where we don't have weeds. When the transplants are a month or two old, we till or dig our garden soil into suitability (destroying all existing weeds in the process) and plant our transplants. We use transplants primarily for other reasons, but one additional benefit is that the transplants have a huge head start over newly germinating weeds or even weeds coming up from fragments of perennial roots. This can greatly mitigate the weed pressure. In fact, if your soil seed bank is so full of weed seeds that gardening is difficult, you may want to shift to using mostly transplants. The combination of transplants and mulch is especially effective in this situation.

By late spring in maritime Oregon the rains have usually stopped and new tilled soil is dry enough so that seed germination is delayed without watering. So for plantings from late spring through the rest of the summer dry season, I frequently make use of selective watering for weed control.

I sometimes use presoaking seeds to create relative delay, for example. The trick depends on two factors. First, it usually requires more water, that is, wetter soil to germinate seeds than to grow the seedlings and plants. This is especially true of big seeds such as peas, beans, or squash, which have to imbibe quite a lot of water before they can germinate. The second factor is that seeds can take up water most efficiently when the soil is compacted around them rather than being loose. If I plant big seeds such as corn, peas, beans, or squash right after tilling and without compressing the soil around them to provide excellent soil–seed contact for transfer of moisture, the seeds usually do not germinate at all until after there has been some rain or irrigation. If I walk on the rows to compress the soil around the seeds, they germinate much faster in the no-rain no-irrigation situation. As do the weed seeds. Often the weeds come up only in the walked-on rows, not between them. That alone can save me an entire weeding of the space between the rows.

But I will still have weeds coming up in the rows. If instead I presoak the seeds, they germinate just as fast as if irrigated or rained on, but without my needing to walk on or otherwise tamp down the soil around the seed. If I presoak the seeds but don't tamp down the soil, only the garden seed has optimal conditions for germinating; the weed seeds do not. So I can give the garden seed the equivalent of a couple of weeks' head start over weed seeds (and, basically, skip one entire weeding) just by presoaking the garden seed. (If you mix some dry soil into the soaked seed you can handle it just as easily as dry seed.)

Presoaking can also be very useful with tiny seeds, especially carrot seed, which takes about three weeks to germinate and which must be kept fully moist the whole time. In my climate, that means watering every day except for the early-spring planting. It's just not something I can usually manage. So instead I mix the carrot seed into a fine weed-free medium such as weed-free compost or peat moss, moisten the mixture, put it in a covered container where it won't dry out, and stir it occasionally until the seed just begins to show the first signs of germination. Then I sow the mixture in my garden rows and water it. The carrots then emerge promptly with no further special attention. It's lots easier to keep a plastic bag with a pint of compost and some carrot seed in it moist for three weeks than to keep the entire surface of the carrot patch optimally moist for three weeks.

When I sow fine seeds in summer I often sow in the bottom of shallow trenches. I make the trenches, sow the seeds in the bottom of the trenches, then rake the seed in lightly (without filling the trenches in). Then I walk down the trenches to compress the soil around the seeds.

Both weeds and garden seed germinate in the bottom of the trenches. But the top layer of the soil dries out much more between the trenches than in them, so I can generally save myself one whole weeding of the space between the rows.

With transplants or with seeds planted in positions with a lot of space between I often water just the planting positions for the first couple of weeks while the seeds are germinating or the transplants are getting established. I already mentioned watering just the planting positions for tomato transplants the first few weeks after planting rather than the entire tomato patch area. An even more extreme example is big viney squash. For example, I might plant the squash in hills 5 feet (1.5 m) apart in the row with rows 10 feet (3 m) apart. Each "hill" is just a line of three seeds spaced 6 inches (15.2 cm) apart that will be thinned to one plant per hill. For the first two weeks after planting I water just the planting positions. (Since I walk on the planted hills after planting, they are depressions that are easy to fill up with water.) Watering just the relevant positions doesn't cost much work and usually saves the equivalent of one weeding of the entire field other than just the planting positions.

When it comes to the removal approach to weeds, timing is critical. The best time to weed is as soon as you see a light fuzz of tiny weeds. The bigger the weeds, the harder they are to remove.

It is a lot easier to hoe weeds from a standing position than to get down on hands and knees and hand-pull them or remove them with hand tools. If you want to maintain a large garden the right hoes and hoeing techniques are critical, as is learning to sharpen your hoes regularly. Alternatively, you may have a more modest garden but also a full-time job and might need to fit all your gardening into a few hours on weekends. You can maintain a surprisingly substantial large garden on just a few hours of work a week once you master hoeing. I discuss hoes, hoe sharpening, and hoeing techniques in the remaining sections of this chapter.

Once you know how to remove weeds, what do you do with them? With tiny weeds cut off at the optimal stage, the weeds stay right where they are cut. With bigger weeds there can be a temptation to remove them. I usually leave cut weeds in the beds so their nutrients are returned to the soil as efficiently and promptly as possible. If the weeds are so big they will interfere with maintaining the garden plants, I toss them in the paths. An exception is large weeds that would re-root if left on the ground. These I put into a bin to dry out and then return them to the garden paths, or I add them to the compost pile to be recycled back into the garden later. I avoid removing weeds from the garden unless there is real reason, though. The more biomass you remove from the garden, the more you need to replace it by growing green manures or cover crops or importing fertility from elsewhere.

Perennial weeds can be a serious problem for organic gardeners. If there are not too many of them, the best tactic is to hand-pull or dig them out when the soil is thoroughly moist so as much of the root comes out as possible. If there are lots of them, you may need to resort to merely tilling or hoeing them down regularly enough so that the roots lose mass instead of gaining more mass

season by season. If you keep after them, ultimately the weeds will be eliminated.

Specific perennial weeds can be the gardener's bane in certain areas. Kudzu is very much not beloved in the Southeast. Here in maritime Oregon our least favorite perennial weed is "bindweed" or "wild morning glory." It grows and spreads as a swath of vines that twine up and cover everything. When rototilled, even the smallest fragment of root can create a new plant, so tilling sometimes just spreads morning glory rather than controlling it. It needs less water than most garden plants, and can even grow and spread all summer with no irrigation, though it grows and spreads faster if watered. Once a field is covered with morning glory, it can be impossible to use for gardening without resorting to herbicides to kill the weed and start over. So it is particularly important if you see morning glory to weed it out immediately. If you use leaves that come from other places for garden fertility or mulching, be vigilant about the possibility of importing wild morning glory seeds. Inspect and weed out any problems promptly.

Garden Woman Meets Pigweed with Attitude

One fine September morning Garden Woman goes out to the garden to pick a few green beans for dinner.

Garden Woman: "Tra la la la la. A pole bean picking I will go . . . *Yipe!* What are *you* doing here!?!" A giant pigweed has sprawled over the whole area. Garden Woman cannot even get to her beans. (See the photo section.)

Giant Pigweed: "Ha ha ha! Puny human! You ignored me when I was little. Now my stalks are half an inch across and solid wood at the base. You'll never get me out of here now! You can kiss your beans good-bye!"

Garden Woman: "Humph! We'll see about that!" She reaches for her secret weapon, her heavy-duty peasant hoe. Hoe in hand she marches back to the garden and assumes her special secret grip with both palms facing downward, her stance such that the hoe is working off to her side, not directly in front of her. Her arm muscles work to lift the hoe, but then simply guide it as it drops, letting gravity do all the work. She orients the hoe so that one of the sharpened flared points enters the plants or ground first, creating a slicing action backed by the full force of the heavy blade. She simply lifts and drops, using arm muscles and swaying a bit with legs and hips, her back comfortably straight the whole time. (See the photo section.) The hoe slices through the woody stems and roots of the pigweed like butter.

Garden Woman: "Take that, Evil Pigweed!" (Thump! Thump! Thump!)

Giant Pigweed: "Ack! Ack! Ack!"

In a just a few minutes the pigweed is history.

Garden Woman: "I know you would just re-root if I left you in the garden, Foul Pigweed. So I'm going to put you in this tub until you are all dried up. Die, Pigweed, die!"

Giant Pigweed: "AARRrrrgggggggggggggg."

Garden Woman: "Hmm. These onions could use a little weeding too." It's just ordinary weeds in the onions—weeds with succulent, not woody, roots and stems. But the spacing between the rows and between the plants is tight. So Garden Woman turns to her Coleman hoe. (See the photo section.) This hoe requires its own secret grip. Garden Woman grasps the hoe handle like a broom, with both thumbs up; the razor-like blade

skims just beneath and almost parallel to the surface of the soil. Garden Woman's back stays comfortably straight while she uses the hoe. In a few short minutes the onion patch is weed-free.

Later that day Garden Woman comes back and picks the pole beans for dinner.

Garden Woman: "Well, it may not look as neat as Eliot Coleman's gardens, but at least I can pick my beans."

The American Square Hoe

Amazingly, the most popular hoe in America, the one with the light square blade set at a wide angle from the handle, is actually not good for hoeing. It would not be good for hoeing even if it were made of good-quality metal that could take and hold an edge, which it usually isn't. The square hoe blade is not heavy enough to do the heavy-duty hacking and digging that is the province of the peasant hoe. And the blade is not set at an acute enough angle to do light weeding either. Weeding with the American square hoe is like cleaning a toilet bowl with a toothbrush.

Basically, you need a minimum of two or three different kinds of hoes. You need one or more light hoes for weeding little or succulent weeds. These hoes depend upon your pushing or pulling them through the soil, not upon lifting and dropping. For these you want good-quality but light-weight blades. Then you need a heavy hoe for bigger weeds and woody material and for removing sod or actual digging. These heavy hoes are lifted and dropped, and you need a heavy blade so gravity does most of the work. For light hoes, I use stirrup hoes for most of my weeding of ordinary weeds in paths and where the spacing permits and

Coleman hoes for weeding in tight spaces, or in the areas right next to plants. When the weeds are bigger or tougher or the soil is compact or actual digging is required I use the peasant hoe.

The American square hoe was invented by American merchandisers who decided that instead of carrying myriad hoes, each excellent for a specific purpose, they would rather carry one hoe that wasn't good for anything. They also decided that the hoe would be made out of cheap metal that could not hold an edge, thus allowing it to be sold at rock-bottom prices. I suppose this makes sense. If you are going to clean a toilet bowl with a toothbrush, the quality and durability of the bristles on the toothbrush is not the outstanding problem. What matters more is how long it will take before you learn that there is such a thing as a toilet bowl brush. Only beginners who don't know better buy the American square hoe, and beginners usually do not sharpen their garden tools anyway. If you are a beginning gardener and are using a square hoe, you can cut your weeding time and effort by at least about 80 percent and your total gardening labor by at least half, most likely, by getting three good hoes of appropriate designs and a file and learning to sharpen your hoes and use them properly. I'm going to describe the three types of hoes I use in detail as well as how to use and sharpen them.

Unless we grew up with parents who gardened, though, most of us do own a square hoe, an artifact of our novice era. We don't bother taking the square hoe to the garden, however. It is left in the garage and used for mixing concrete. Or the handle is removed and used when we need a chunk of dowel to secure windows. The square hoe is not completely useless in gardening, however. If you have bags of minerals you want to mix before spreading, you can dump them all on the driveway and use the square hoe to move them around and mix them. For moving stuff around the square hoe is actually hard to beat, I've found. Its dull blade has just the right angle and weight and dimensions for that job. Just don't try to use it for hoeing.

Buying, Using, and Sharpening the Peasant Hoe

If I could have just one garden tool, it would be my heavy-duty peasant hoe. I bought mine from Red Pig Garden Tools, which as far as I know is the only company that sells either the blade style I prefer or any heavy-duty hoe with a long enough handle to use my special secret grip that allows hoeing standing straight instead of bending over. (See www.redpigtools.com. Look under Hoeing, then click on "Eye Hoe (Peasant Hoe)." Or call them at 503-663-9404.) *Eye hoe* refers to the fact that there is a hole in the metal ring part of the blade through which you insert a screw to help stabilize the heavy blade on the handle. My peasant hoe is the one with the 2-pound (0.9 kg) blade that is 7.75 inches (19.7 cm) wide and 9 inches (22.9 cm) long. The blade flares out at the corners instead of the sides being parallel. This style is made for sharpening the sides of the blade at the points as well as the blade, allowing a slicing as well as chopping action when the hoe is canted so it enters the weed stems or ground point-first. The slice-chop gives a much more effective stroke for the same amount of energy as a pure chop. Get the longest handle, absolutely no less than 60 inches (1.5 m). At the moment the Red Pig Garden Tools peasant hoe costs $45 plus shipping. Shipping raises the cost a good bit, as the hoe is heavy; in addition, shipping longer handles is expensive.

I would actually prefer a handle a little bit longer than the 60-inch (1.5 m) handle, the longest one currently available for the peasant hoe. The 60-inch handle is just barely long enough to use my special grip and stance given my height (5 foot 8/173 cm). Anyone much taller than I will absolutely need a longer handle to use the secret grip and hoe without back strain. I've been talking with Bob Denman at Red Pig, and he has agreed to look into adding a longer handle to his standard options for the peasant hoe.

I use my peasant hoe for hacking down big established or woody weeds, taking out blackberry vines or brambles, removing the sod from areas to be turned into garden beds, breaking up heavy clods, digging planting holes, hilling up potatoes, knocking down cornstalks or other heavy crop residues, and reworking the ground lightly between succession crops in lieu of tilling between crops during the season. There are other hoes that do the lighter weeding of small tender weeds much more easily or are better for using in tight places, but only the peasant hoe can handle big, tough weeds, take on woody stems or brambles, or substitute for a shovel, plow, or tiller.

When most people grab a heavy hoe for the first time they intuitively grip it like a baseball bat, with the two palms facing each other. If you use this grip, the hoe's working position is directly in front of you, and your back's working position is stooped over and painful. In addition, you will be lifting the heavy hoe blade with your back and using your back to power the down stroke. It's a back-killing way to use any hoe, especially a heavy

hoe. However, it is the only method possible with a short-handled hoe. Instead, get a hoe with a long enough handle. Then grip it with both palms down, with the blade working off to the side, your back comfortably straight. (See the photo section.) Your thumbs can curve around the handle in the direction opposite to the rest of your fingers if you like. I prefer to keep my thumbs on the same side of the handle as the rest of my fingers, as that grip holds the hoe just as well, and allows me to work for hours without needing gloves. (With the thumb curled around opposite my fingers, I get blisters if I hoe for very long without gloves.) You don't bend or exert power from your back at all if you are using the hoe properly. Instead, you use arm muscles and a bit of swaying from hips and legs to lift the blade as high as is needed to get the amount of power you want for the kind of work you are doing. Then you drop the blade, letting gravity do all the work, your arm muscles simply guiding and aiming.

No tool, not even the mighty peasant hoe, is much good when it is dull. I have a bench grinder but do not use it for sharpening garden tools. Power grinders are too inaccurate for the job, and it's too easy to grind off much more metal than necessary. And sharpening most hoes with a file is so easy and fast, I can do it in much less time than it takes to get out and hook up a bench grinder. In addition, for optimal ease of work, the hoes are sharpened frequently and regularly right in the field where there is usually no electricity for power tools. The ideal tool for hoe sharpening is a metal file mounted on a wood or plastic handle. I use flat files that are about 8 or 9 mil as well as files that are flat on one side and rounded on the other for sharpening both hoes and shovels. There is a very nice "Gardener's File" with an integral handle at www.easydigging.com for $6.50. Files themselves need to be replaced periodically as they wear down.

I sharpen my peasant hoe about every half hour of working with it. I like to take a break about that often anyway, sit down and contemplate the world a few minutes, or make notes in my notebook while drinking occasional swigs of water. Sharpening the hoe fits nicely into these breaks. It takes about two minutes. I carry the metal file in my van near the back bumper where it is handy, and sharpen the hoe by propping it up on the van's rear bumper.

I sharpen the peasant hoe to an edge of about 30 degrees. (About one-third of a right angle is easy to see and remember.) I align my wrist and arm straight and put my weight behind the strokes. If you haven't propped the blade so that it is solid and motion-free as you sharpen, or if you have your wrist angled off so you can't put real force into the stroke, you can sharpen quite a while without achieving much. If you are doing it right, though, the powdered metal sprays off obviously with each stroke. I sharpen just one side of the blade except for very light strokes from the off side to remove feathering. I might, for example, use three to five full-weight strokes on the main side, then four very light (barely touching, almost caressing) strokes on alternating sides to remove the feathering. (I don't want to change the shape of the off side, just remove the feathering, the fine delicate strips of metal that bend over the blade when it is sharpened.) (The Gardener's File sold by www.easydigger.com actually has one coarse side for use in sharpening the primary edge and a finer side to remove the feathering.) I use strokes nearly parallel to the edge or that angle slightly from blade body to edge. (The opposite orienta-

tion of stroke that is more common when sharpening knives, from edge to blade body, will put nicks in the edge if used with these metal files, so it actually ruins and dulls the edge.) I use the same pattern to sharpen the lower inch or two of the sides of the peasant hoe as well.

Buying, Using, and Sharpening the Coleman Hoe

The Coleman hoe, also called a collinear or collineal hoe, was designed by Eliot Coleman, whose books are full of images of perfect weed-free garden beds that no part of any garden of mine has ever even approached. The Coleman hoe is my favorite tool for thinning or weeding in tight places or close to plants. With the Coleman hoe you can work right up against plant stems and underneath spreading foliage. Many mail-order tool and seed companies carry Coleman hoes, but most offer only some of the sizes and models available. I get my Coleman hoes from Johnny's Selected Seeds, which carries the full selection of sizes and models, and at competitive prices.

You can get standard or replaceable blade models. I prefer the standard rather than the replaceable blade model because the part that enters the ground is more streamlined and creates less drag than with the replaceable blade model. The blade widths available for the standard model are 3.75 inch, 5.5 inch, and 7 inch (9.5 cm, 14 cm, and 17.8 cm). I prefer the smallest size of Coleman hoe, as it is best for thinning and weeding tightly spaced plants. The wider models are better for thinning and weeding in bigger spaces, or for weeding right next to bigger, more widely spaced plants with spreading foliage. Johnny's

Selected Seeds' current price on the fixed-blade Coleman hoes is about $38, whatever the size.

The secret grip for the Coleman hoe is with both thumbs up. You can wrap your thumbs around the handle or let them rest on the handle pointing upward (away from the blade end). The blade works directly in front of you and quite near your feet. You can use short strokes, starting each by placing the blade on the ground. Alternatively, when soil conditions are right, you can simply move the blade along the soil just above or just under the soil surface and take out everything.

I use the same style of file sharpening on a Coleman hoe as on the peasant hoe except that the edge is a little more acute, and the sides are not sharpened at all.

Stirrup Hoes

Stirrup or oscillating hoes have a thin curved blade that is held in such a way that only the blade enters the ground. The oscillating action allows you to weed on both the push and pull strokes. There are many cheap worthless versions of stirrup hoes in most big-box stores and garden centers. Don't bother with them. They are poorly designed and made of metal that won't hold an edge.

My stirrup hoes are the Swiss-made Glasers. I use them for weeding in paths and between rows where spacing isn't too tight. I also thin with them where the plants are not too close together. The stirrup hoe is not good for working right near the stems of plants since the side edges are sharp. Nor is it good for dealing with heavy weeds.

I have to admit that until recently I hated stirrup hoes and did not use them, because the stan-dard way of using them was so hard on my back. Only years after I discovered my secret grip for using the peasant hoe did it occur to me to try that same, counterintuitive, both-palms-facing-down grip on a stirrup hoe. That grip made all the difference. The Glaser stirrup hoe is now the hoe that does most of my weeding, with the peasant hoe used just for heavy weeding, weeding compacted soil, hacking, and digging, and the Coleman hoe used just for thinning and hoeing in tight spaces. As with the peasant hoe, when gripped with the secret grip, the stirrup hoe works out to your side, not in front of you.

Johnny's Selected Seeds has the full variety of sizes of Glaser stirrup hoes, and sells them with longer handles than some other companies. However, their 59- or 60-inch (1.50 or 1.52 m) handle is still really a bit short for using opti-mally with my secret grip. In addition, I find myself sometimes putting the palm of my hand against the end to get my arm and body force directly behind the hoe, which is really hard on the hand. Johnny's makes a knob that attaches to the end of the stirrup hoe. I ordered one, but never attached it. The knob has to be drilled, and the end of the handle has to be cut off and drilled accurately lengthwise down the handle. It's a job that requires more woodworking equipment and skills than I have. So, instead, I took the 72-inch (183 cm) pistol-grip handle off my diamond hoe, reshaped the end a bit, and mounted it on a Glaser hoe. That's now what I use for most of my light hoeing. The longer handle allows me to use my secret grip along the length. As an alternative, I can use the pistol grip and hoe one-handed with either hand. This gives me a delightful range of options that allows me to swap off and use differ-ent arm muscles and hoeing styles, all of which allow me to hoe standing up straight without using my back.

The Glaser stirrup hoes come in three widths (3.25 inch, 5 inch, and 7 inch; 9.53 cm, 12.7 cm, 17.8 cm). I have the small one and two of the medium-sized and mostly use the mediums. The smaller size is better for thinning and weeding between rows that have very narrow spacing. The larger two sizes are better for weeding paths and between widely spaced plants. Except when the weeds have become too tough or woody, I do all my weeding of the paths between rows with either these stirrup hoes or with a stirrup hoe attachment on a wheel hoe. (See the section on wheel hoes below.)

The stirrup hoe seems to be more sensitive to moisture level in the ground than most hoes. It's far easier to hoe with a stirrup hoe when the ground is just barely dried out on top and a little moister than when hoeing with most other hoes. When the ground is dry, the hoeing is much harder. The stirrup hoe is best with loose soil. If the soil has become compacted, I reach for my peasant hoe.

Sharpening the peasant hoe and the Coleman hoe is very fast and easy. Sharpening the Glaser stirrup hoe is a colossal pain. I have three in part because I don't sharpen them in the field. The basic process is the same as for the other hoes, but you have to find just the right spot to prop the hoe so as to be able to keep the head from oscillating as you sharpen. In addition, you need to sharpen to a more acute angle, and sharpening the convex bends is *really* a pain. However, once I mastered the stirrup hoe I found that I no longer needed a rototiller (and a guy to run it) to keep my paths and garden weeded. I could do the half-acre (0.2-hectare) garden myself. So I grit my teeth and put up with the sharpening. It takes me about ten minutes or so to sharpen a Glaser stirrup hoe.

Wheel Hoes

A wheel hoe is a support structure mounted on a wheel to which you attach a stirrup hoe or other tools of the same kinds as you use on hand tools. You can work much more effortlessly and efficiently with a wheel hoe because the wheel and mount do the work of holding the tool up and at the correct working angle to the soil. You need only push the wheeled tool along, not hold it up. Furthermore, with the tool supported by the wheel, you use your body weight and leg muscles as well as your arms to move the tool along, which is much easier than using just the muscles in your arms. Many people use a rototiller set at a shallow depth to weed the aisles or larger empty spaces in the bigger garden. Unless the garden is bigger than 2 acres (0.8 hectare), however, the wheel hoe might be the better tool for the job.

I detest rototillers. It is actually much easier on my body to weed the same amount of space with a wheel hoe than with a rototiller. A wheel hoe is quiet and pleasant to use compared with a rototiller. A rototiller is so hard on my back that I am unwilling to use one at all. I need to have a guy along with the rototiller to run it. Most rototillers don't come with a guy. He is separate.

Rototillers also seem to break down regularly, usually exactly whenever they are needed most. The wheel hoe doesn't break down. These days I handle all my weeding of my half-acre (0.2-hectare) home garden with a wheel hoe and hand tools.

The rototiller not only is noisy and hard on the body, but also restricts and dictates your gardening patterns in a way that ends up wasting space and requiring more space for the same amount of crops—hence a bigger garden than you would need otherwise. I can grow about as much food on half an acre weeded with a wheel hoe and hand tools as on an acre spaced so that weeding between rows can be done with a rototiller. If you have a 30-inch (76 cm) rototiller, for example, you will usually need to space all your rows at least 36 inches (91 cm) apart. Those plants that require much less space either cause wasted space or end up in beds hoed by hand. The wheel hoe can handle rows that are just a few inches wide if need be, so it allows for much more flexibility in row spacing and better use of the garden space.

In addition, rototillers are a big pain to turn around at the end of rows, so much so that if you weed rows with a rototiller you will probably need to design your garden as long straight rows. However, there are often good reasons to plant in blocks rather than in long rows. Wheel hoes are light and easy to turn around at the ends of rows. They allow you to configure your garden in a greater variety of ways, with rows of various lengths. In addition, you can easily hoe curved rows with a wheel hoe. I appreciate this greater versatility because my garden is on sloped land. I like to be able to change the orientation of the rows to parallel the slope of the land.

The rototiller can be used for breaking ground. A wheel hoe can't. But a walk-behind rototiller is so poor at breaking ground that most people with bigger gardens, even those who own rototillers, usually call the tractor guy to do the primary tilling and use their rototiller just for weeding. But that job doesn't require a rototiller. The wheel hoe might be the better choice. It may be better even for much of the weeding on larger market gardens.

A final factor—I don't know anyone who enjoys using a rototiller. Most people seem to enjoy using wheel hoes. They are fast, quiet, and efficient—appropriate technology at its most appropriate and finest.

There are two basic designs of wheel hoes, the high-wheel hoe and the low-wheel hoe. The high-wheel designs are cheaper but harder to use, and they don't transfer the force from your body nearly as efficiently as the modern low-wheel designs. The high-wheel variant is outmoded technology. Get a low-wheel model. The Glaser, the Valley Oak, and the Hoss are all excellent choices. The Glaser is Swiss-made and is available from Johnny's Selected Seeds and Peaceful Valley Farm Supply. The Valley Oak is made in California (www.valleyoaktool.com) and is available from the manufacturer, from Fedco Seeds, and from Peaceful Valley Farm Supply. Hoss is made in Georgia (www.hosstools.com) and is available from their website and from www.easydigging .com in the United States and from William Dam Seeds in Canada. The Hoss website has numerous videos showing the use of their wheel hoes with various attachments. The www.easydigging.com website has a very complete set of videos showing how to use all the different configurations and implements for various purposes and with various crops. All three brands allow you to adjust the handles for the height of the user and offer oscillating (stirrup-style) hoes in a number of sizes, a tine cultivator, and a furrower as attachments. The Hoss has a bigger repertoire of attachments than the others, though, and their attachments are better designed and much more versatile. The

Hoss wheel hoe and attachments are also considerably less expensive than the others.

In addition to the oscillating hoes, tine cultivators, and furrower, the Hoss has sweeps as well as a fully adjustable six-disk harrow. Their furrower is actually a left and right plow mounted together. You can use just one to plow or hill up soil to just one side. In addition, you can turn the plows around and spread them out on the mounting frame and use that configuration to cover up the furrows. The other brands do not give you attachments that allow you to plow, hill up just to one side, or cover furrows. The Hoss tine cultivator is also more versatile than those of the other brands. You buy separate tines and configure them on the frame into cultivators of one to five tines with whatever spacing between the tines you want. In addition, the tines themselves are better designed.

All the Hoss implements except their seeder can also be used on the Glaser wheel hoe. If you already have a Glaser with one or two implements, you might want to switch to Hoss for additional implements.

The Hoss wheel hoe is available in one- and two-wheel models. The latter is necessary for setups involving two hoes that straddle the row. Otherwise, the one-wheel version is more maneuverable. (I don't aspire to row straddling. You have to have much straighter, more uniform rows than I have and be much more attentive when weeding than I like to be for row straddling to be workable.)

The Hoss also comes with your choice of either standard plow-style or pistol grips. The pistol grips allow you to better transfer force from your arms and body weight to the handles because your wrists are straight when using the tool. The plow-style grips give you somewhat better lateral control, but your wrists are bent as you use the tool. The pistol grips are more likely to give you blisters if you don't wear gloves. The plow-style grips are more likely to cause fatigue or soreness in your wrists. I bought a Glaser initially because it was the only wheel hoe I had ever seen or used. I've subsequently bought a Hoss with a complete set of implements. I want the Hoss for its ability to cover furrows as well as make them and for the six-disk harrow for making fine seedbeds. I plan to use the Hoss for a season or two and then post an article on my website evaluating and comparing the Hoss and the Glaser.

I have never seen a Valley Oak wheel hoe. From the pictures it seems to me that the way the handles attach to the frame is not as good a design and would give you less control than you get with the Glaser and the Hoss.

An 8-inch (20.3 cm) stirrup hoe for basic weeding is the first implement most people buy and often comes with the wheel hoe frame. It's best for general use. Five-inch (12.7 cm) and 12-inch (30.5 cm) stirrup blades are also available. The bigger size is most suitable for very soft or sandy ground. The smaller size is most useful for weeding where the spacing between the rows is tight. I got an 8-inch stirrup hoe attachment for doing most of my weeding between rows and a furrower attachment for making furrows for planting potatoes and bigger-seeded crops like corn and beans.

A Glaser wheel hoe with the basic three attachments (oscillating hoe, three- or four-tine cultivator, and furrower or furrower/plow set) costs about $560 at Johnny's Selected Seeds. The Valley Oak wheel hoe with those three attachments costs about $440 at Fedco Seeds, which is actually a bit less expensive than it is direct from the manufacturer's website. The Hoss wheel hoe with those three attachments in a special package deal at the www.easydigging.com website is $227. In addition, the furrower also gives you plowing, hilling, and furrow-covering capability. The prices of the Hoss at www.easydigging.com are about the same or a bit less expensive than from the manufacturer's website. And, in addition, the package includes a durable oil finish on the handles and professional sharpening of the tools.

When the weeds are tiny and tender, you can weed with a wheel hoe with a stirrup hoe attachment by just walking straight down the row about as fast as you would walk normally, pushing the wheel hoe in front of you. Ideally, you keep your garden weed-free by making such quick, almost effortless passes over it frequently. If the weeds are bigger or the soil is heavier, you use forward and back motions. You push forward a couple of feet, then draw back a foot, then push forward again. Even with the push–pull motion you can hoe so fast it is almost just like a slow walk down the row. See the YouTube video of Sam Eglington demonstrating weeding with his Glaser wheel hoe and the videos at www.easydigging.com to get an idea of just how easy and fast weeding is with these tools.

Electric Wheel Hoe and Electric Tiller

If you have a garden that is a bit too big to weed comfortably with a wheel hoe, there is a new level of appropriate technology that has been developed by Carts and Tools—the Solus electric wheel hoe and the small but powerful Tillie electric tiller. (See cartsandtools.com.)

The Solus electric wheel hoe is basically a Glaser wheel hoe with a powerful battery set up to drive a wheel and a 12-inch (30.5 cm) stirrup blade. It takes all the accessories available with Glaser wheel hoes and will presumably also take all the Hoss implements except the seeder. My colleague Paul Harcombe, a grower for some of the seed crops I breed and sell through Fertile Valley Seeds, has recently switched from the Glaser wheel hoe to the Carts and Tools electric wheel hoe for nearly all the hoeing between the rows of the 1.5 acres (0.6 hectare) of corn, beans, and squash he grows for me.

Paul tells me that the electric wheel hoe is just a little faster than the standard wheel hoe, but is *much* easier on the body. You just need to use your muscles to guide the hoe rather than pushing it through the ground. In addition, the electric wheel hoe can dig deeper if desired, and can hoe much coarser ground than the standard wheel hoe. The electric wheel hoe costs about $800.

The Carts and Tools electric tiller, called the Tillie, was initially designed as a way to plant and hill up potatoes. It too is battery-operated. It can do just about everything except primary tillage. It can be used to make and cover furrows, form raised beds, hill up potatoes, cultivate, and weed. The Tillie also costs about $800.

道

Be detached. Work with joy, without caring for the achievement.
Travel with joy, without focusing on the destination.

On disaster good fortune perches. Beneath good fortune disaster crouches.
Straight becomes crooked, which becomes straight again. Good gives rise to evil,
which gives rise to good again. There is no beginning or ending.

The bright way may seem dark. The straight way may seem crooked.
The smooth way may seem rough. The way forward may seem to go backward.
True love may seem heartless. True integrity may seem dishonest.
True fullness may seem empty. And even the eternal may seem to change.

Success is as treacherous as failure. Praise is as useless as blame.
Fortune, fame, and favor are as dangerous as disgrace.
High status and position can bring you great affliction.

You can lose by gaining. You can gain by losing.

Trapped in desires, you see only your desires. Trapped in desires, you see surfaces.
Free from desires, you see the subtleties. Free from desires, you experience the mystery.

To know that you know not is sanity. To know not that you know
not is sickness. The Sage is her own healer. She cures herself of false knowing.

The Sage practices non-knowing. She knows that she does not know
what is good and what is evil. So she cleanses her mind of desire for particular
ends, and accepts everything, and lets everything happen, and takes joy
in everything, and uses everything that comes her way.

CHAPTER NINE

Non-Knowing—Squash

Adventures in Ignorance. The Perfect Polyculture—Squash and Overwintering Kale. 'Candystick Dessert Delicata' Squash. 'Lofthouse Landrace Moschata' Squash. Apologizing to a Squash. Butternut Squash Cookery. Planting by the Moon. Talking to Your Plants. True Understanding.

Once upon a time in a village in ancient China there was an old man who lived alone with his son. They were very poor. They had just a small plot of land outside the village to grow rice and vegetables and a rude hut to live in. But they also had a good mare. It was the son's pride and joy, and their only possession of value.

One day the mare ran away. The old man's friends came to him and commiserated. "What a wonderful mare that was!" they said. "What bad fortune that she ran off!"

"Who can tell?" the old man said.

Two weeks later the mare returned accompanied by a fine barbarian stallion. Friends and neighbors all came around and congratulated the old man. "Now you have your mare back, and that stallion is as fine as any in the land. What a stroke of good fortune!"

"Who can tell?" the old man said.

Two weeks later the son fell off the stallion while riding and broke his leg. Friends of the old man came to him to express their sympathy. "It's too bad your son broke his leg. And right before the planting season, too. What bad luck!"

"Who can tell?" the old man said.

Two weeks later, war came to the land, and all able-bodied young men were drafted. The troop that contained the men from the village was at the front in a bloody engagement, and the entire troop was lost. All the men from the village died in battle.

The young man with the broken leg stayed home. His leg healed. He and his father bred many fine horses and tended their fields.

Adventures in Ignorance

I apparently have had celiac disease my entire life, but it did not start causing me obvious problems until about the same time as I began breeding plants seriously. Celiac disease is an autoimmune disease in which, in response to a specific protein in wheat and related grains, your immune system attacks the lining of your digestive tract, especially of the small intestine. That autoimmune reaction trashes and destroys many of the villi that are responsible for absorbing food. People with celiac disease may have digestive system pain and symptoms of various sorts, but whatever else, they don't absorb food nutrients very well, especially fats and lipids, lipid-soluble vitamins, and certain other vitamins and essential nutrients. So celiacs are usually seriously vitamin-deficient in addition to whatever obvious symptoms they have. The solution is simple: Don't eat anything containing any form of wheat or any of its close relatives (spelt, emmer, rye, barley, triticale).

Long before my celiac disease began manifesting as obvious symptoms, I was undoubtedly deficient in many vitamins. My interest in gardening in general as well as my development of the eat-all greens garden growing pattern undoubtedly resulted partially from vitamin deficiencies. I craved greens. I would cook up a pound or more of greens and dress them with something delicious. Then, my hand trembling with need so violently it was hard to hold the fork, I would bolt the greens down with the desperation of a drug addict overdue for a fix. It was very obvious that these greens had something I was missing. Taking vitamin pills did not affect the situation. Either I needed something not in the pills, or was not able to assimilate what I needed in pill form. Raw greens, even of the same kinds as were so satisfying cooked, also didn't help. I apparently could not digest the raw greens well enough to assimilate what I needed. I needed cooked greens. And it was simply not possible to get enough greens of the kinds I craved most—kale, mustard greens, and other high-nutrient cooking greens—without growing them myself. It was difficult to grow as much of these greens as I seemed to need on the small amounts of land I had available at that time. I needed to be able to grow more greens on less space, and more easily. This personal dietary problem was undoubtedly the basic drive behind the research and work I put into developing the eat-all greens garden approach.

Long before I knew that I had celiac disease and needed to avoid most of the wheat-containing staple foods that are readily available commercially, I found that my plant breeding attention was drawn primarily to breeding backyard staple crops, gourmet-quality varieties of winter squash and corn for cornbread and polenta as well as dry beans. My homegrown staples were so good I tended to not eat wheat or wheat-containing commercial staples when I had the homegrown crops available. So growing my own gourmet-quality staples helped my health even before I knew I should be avoiding wheat. On a subconscious level I probably recognized this.

People with celiac disease have leaky guts, so they often develop other allergies and other food allergies. I developed asthma and started experiencing frequent life-threatening attacks. Developing asthma seemed like a bad thing at the time, but it turned out to be a lifesaver. I began experiencing serious obvious immediate allergic reactions to wheat—asthmatic attacks and symptoms of anaphylactic shock. So I quit eating wheat because of the allergic reactions. Only after eliminating wheat for allergy reasons did it become obvious (from occasional accidental ingestions of small amounts of wheat-containing products) that wheat caused not only the allergy symptoms but also all the digestive system symptoms I had been experiencing. I had seen a doctor much earlier about what were actually symptoms of celiac disease, but got nowhere (the typical situation for those with celiac disease). The asthma and anaphylactic shock reactions served as an effective bioassay to tell me what the problem was.

One other aspect that became obvious after I eliminated wheat from my diet was that I was also severely deficient in omega-3 fatty acids. This is not surprising, given that people with celiac disease have an especially hard time absorbing lipids and fats, and the omega-3 fatty acids are lipids. In addition, omega-3 fatty acids have been almost eliminated from the standard American agribusiness diet by the replacement of grass-fed meat and animal products with those from animals that are grain-fed. Omega-3 deficiency is known to cause or exacerbate allergies, asthma, hay fever, and other immune and autoimmune disease. So I began eating free-range duck eggs and grass-fed meat. (Cod liver oil also works.) After that the asthma went away. And without wheat, all the digestive system problems also vanished. So did the hay fever I had had every spring since childhood, and the general difficulty breathing through my nose, which I had also had since childhood. I had always thought those problems were just part of me. In actuality, both probably represented dietary deficiency for omega-3 fatty acids plus exposure to wheat, which my system cannot tolerate. Celiac disease, asthma, hay fever, stuffed-up nose, and hives are all immune system overreactions, not natural conditions, and can be caused or mitigated by diet.

Now I don't eat wheat. And I breed gourmet-quality backyard staple food crops such as squash and cornbread corn and polenta corn and beans that are so good they make me not miss the wheat. And even though I'm not so desperate about greens as I was when I had active celiac disease, I still love them and relish eating large amounts of them. With no wheat in my diet, all symptoms of the celiac disease are gone. So is the asthma. So are the hay fever and stuffy nose I had had since childhood. What remains is the way that my celiac disease has contributed, and continues to contribute, to the focus and originality of my garden research and plant breeding work.

The Perfect Polyculture— Squash and Overwintering Kale

I love 'Russian Hunger Gap' kale. It is very vigorous, has excellent flavor both cooked and raw, and is much more cold-hardy and better at overwintering in my region than even the kales especially noted for overwintering ability (such as 'White Russian' and 'Red Russian'). In addition, 'Russian Hunger Gap' bolts later in spring than other kales, so continues to produce greens through the late-spring "hunger gap" when other overwintered greens have bolted and the spring-planted greens are not yet available. 'Russian Hunger Gap' was one of the kales brought to North America from a collecting expedition to Europe by Andrew Still and Sarah Kleeger of the Seed Ambassadors Project; it is sold through the associated company Adaptive Seeds.

Naturally, I wanted to grow a seed crop of 'Russian Hunger Gap'. So I sowed and grew out and overwintered about two hundred plants that grew to 6-foot (1.8 m) tall bushes covered with pods in late spring. Then the LBBs moved in. Little Brown Birds. Every kale bush was alive with the rustlings of dozens of tiny birds stripping the seeds from the pods and devouring them. I tried cutting and drying some branches just before the stage at which the birds harvested, but ended up with shriveled seed. The birds were eating the green succulent seed, not the dried seed. They ate the seed at a stage that was way too early for me to harvest. I garden right next to a wetland with lots of birds. They took all the seed except for the smaller, later-finishing pods. I guess those didn't meet their standards. They didn't meet mine either. I want the biggest, fattest seed from the earlier and main-season pods, not the smallest seed. So I tilled under the entire patch and started thinking in terms of subcontracting with someone else who didn't live next to a wetland for my 'Russian Hunger Gap' kale growout. I was not very happy about losing my kale seed crop or being unable to grow a kale seed crop in my home garden area. But I could still grow the crop for eating.

After I had the failed kale seed crop tilled under I planted my squash patch in the area. When weeds came up they were, delightfully, mostly 'Russian Hunger Gap' kale. (See the photo section.) Wow. I had a squash patch in which virtually every weed of size was a plant of my favorite kale variety. So instead of weeding, I let the kale grow for a while.

The kale overgrew the squash initially. I harvested the kale for the table from the areas immediately around the squash planting positions, weeding the young squash at the same time. Later, I continued harvesting whole young kale plants as well as weeding them out within the squash rows. Instead of planting kale for overwintering in a new patch by itself in July as I did usually, I just left big swaths of kale between the rows in the squash patch. The squash were not going to be needing that area midway between the rows until August. Maybe the squash and kale could share. If not, I'd let them fight it out and see what happened. In most polycultures two different interplanted species compete for both air space (sun) and ground space (nutrients and water). The squash and kale root systems were in completely different areas. The two species would be competing only for air space and only in August and September. By October the squash would all die back and the kale could then have the patch until the following spring.

By August the squash vines had filled up the space between the rows and were overgrowing the kale. But the kale plants managed to compete well enough to hold their own. They were in the space first, so they were good-sized plants before the squash vines got to them. Most of the kale plants managed to keep at least some of their leaves above the squash vines. In addition kale, like most greens, is tolerant of shade. My subjective impression was that the kale looked much happier growing in the squash patch jungle than it ever had growing alone in a mostly bare-soil monoculture. Whether the kale actually did better than in years when I planted it separately I don't know. What I do know is that I had more kale plants than ever before, and they cost me no space or extra watering at all beyond what I would do normally with my squash. My ordinary way of planting kale required an entire separate patch, meaning extra space and extra watering. Instead, all the kale cost me was a little extra time and meticulousness in weeding around the kale plants instead of the somewhat faster weeding work of weeding empty spaces between rows.

Since I plant many kinds of squash and trial new kinds all the time in my home garden, there are always occasional hills of some varieties that fail to thrive. This year I hoed these under and let the kale have the space. Where the hill was close to the edge of the patch and more easily accessible, I planted small patches of later successional plantings of 'Oregon Giant Sugar' pod pea. They thrived too.

By late summer the squash patch had turned into a glorious jungle of squash and kale with occasional clumps of peas in spots here and there on the edges. (See the photo section.) It was completely impossible to get into the patch. But I didn't need to. Once the squash vines start running, they shade out the weeds, and weeding is both impossible and unnecessary. The squash were winter squash, to be harvested only at the end of the season. The kale was for use in late fall, winter, and early spring, not summer. The hoses and sprinklers for the squash patch, once set, are not moved until the end of the season. (They can't be moved once the squash vines have begun to run.) And the pea patches were all on the edges where I could still reach them to pick the peas.

In late fall after cold weather blackened, killed, and flattened the squash vines, it was easy to see and harvest the winter squash and pumpkins. The kale inherited the space, and grew and produced happily all winter and spring.

I got a huge crop of overwintering kale plus several summer and fall successions of pea plantings all out of the squash patch. I needed no extra space at all for the kale crop and had a much larger crop than if I had needed extra space for it. There is one disadvantage in that multicropping the squash with overwintering kale means I can't till the squash vines under in fall. That might potentially lead to disease problems. But it hasn't so far. And I am trying to move away from fall and spring (twice per year) tilling to just spring (once per year) tilling anyway.

These days I plant all my fall-harvested and overwintering brassica crops as polycultures, each planted between the rows in the squash patch. I simply sow the seed in July, the ordinary time for planting overwintering brassicas in my area. Where the squash are summer squash and I need to get to them to harvest during the season I leave some space for paths. And I plant all the later successions of peas in small patches near the edge of the squash patch. What started as a happy accident has turned into practice.

'Candystick Dessert Delicata' Squash

There is a large chapter in *The Resilient Gardener* on growing and using summer and winter squash, so in this book I limit myself to discussing things I have learned about squash more recently. 'Candystick Dessert Delicata' is a new variety. Here is its story.

There are few winter squash varieties in existence that can match the sweetness and intensity of flavor of the delicata types. These varieties have a flavor that is distinctly different from that of all other squash and pumpkins. The fruits are generally loaf-shaped with longitudinal green stripes against a cream or tan background (depending on variety) except for 'Sweet Dumpling', which is roundish and flattened or punched in around the stem end. The green and cream delicata varieties are 'Delicata', 'Delicata—JS', 'Delicata—Zeppelin', 'Cornell Bush Delicata', and 'Sweet Dumpling'.

The green-and-tan-striped varieties, 'Sugar Loaf' and 'Honeyboat', were both bred by Jim Baggett out of a single plant in a field of 'Delicata' that had the tan-and-green-striped fruits. Tan delicatas are like turbocharged delicatas. They are even sweeter and more intensely flavorful than any of the cream lines. Unfortunately, the commercial 'Sugar Loaf' has lost quality and vigor to the point where it is no longer worth growing. Fortunately, the variety was maintained independently and further selected by Mike Hessel, who, as a graduate student with Jim Baggett, had been involved with 'Sugar Loaf' from the beginning. Mike's line, 'Sugar Loaf—Hessel' is the best line of 'Sugar Loaf' available. (The seed is sold by Nichols Garden Nursery.)

Unfortunately, none of these delicata varieties produce fruits bigger than about 1.5 pounds/680 g. In addition, the flesh is thin, typically no more than about half an inch (1.3 cm) thick. These squash are often referred to as just the right size for a serving for one person. Apparently many people consider only a few bites of vegetable a serving. I like much bigger servings of vegetables; I need two or three delicatas that size to make what I consider a serving. This means a lot of labor cutting and cleaning squash per serving. I wanted bigger delicatas, or delicatas with thicker flesh, or both. That is, I wanted much more edible food per squash. So it was time to breed a better delicata.

One logical approach would have been to cross my favorite delicata variety of that era, 'Sugar Loaf—Hessel', to another *Cucurbita pepo* variety that was much bigger but also of good culinary quality, then recover delicata flavor and quality combined with bigger size by recombination and selection in future generations. Unfortunately, all of the *C. pepo* varieties of good culinary quality are relatively small. So if I crossed the 'Sugar Loaf—Hessel' to something much bigger, it would have to be to something much bigger of poor quality, for example, one of the big, watery, coarse-fleshed, flavorless Halloween pumpkins. However, generally all the poor quality characteristics tend to be dominant in crosses. And many characteristics are involved in fruit quality with many genes associated with each characteristic. That is, a cross between a delicata and a big pumpkin with poor eating quality would involve dozens, perhaps even hundreds of genes affecting fruit size and quality, with nearly all the genes for poor quality being dominant. With so many dominant genes for poor quality in a cross, I would likely have to grow hundreds or even thousands of plants for generations in order to recover a plant with good delicata flavor and quality. I wasn't up for it. I decided to take a more modest approach. I would cross different delicata varieties to each other. That approach would be less likely to give me a really big delicata, but more likely to give me prime-quality fruits at every stage of the breeding project. This would allow us to do the breeding project with little or no more space or number of plants than we might usually devote to growing delicatas for the table anyway. And just a modest increase in fruit size or flesh thickness would make a big difference in the amount of edible food per fruit.

How could I possibly end up with bigger fruits or thicker flesh by crossing two varieties that both have small fruits and thin flesh? The trick is that the two varieties might have different genes for size or flesh thickness. Usually, characteristics such as fruit size or flesh thickness are quantitatively inherited, that is, many genes contribute, often dozens. But to simplify, let's say that all the genes symbolized by letters in this example affect fruit size, with the capital letters representing the alleles (variants of the genes) that contribute larger fruit size and all the lowercase letters representing alleles that contribute smaller size. Now suppose that one proposed parent variety is genetically *AAbbCCdd* and another is *aaBBccDD*. Both varieties have fruits the same size, but they have different genes affecting size. That is, the two varieties have the same size but for different reasons. The F_1 hybrid between the two varieties would be *AaBbCcDd*. (The hybrid may or may not have fruits bigger than the parents depending on which genes are dominant.) From that *AaBbCcDd* hybrid, by selecting in further generations, I should be able to derive a new variety of genotype *AABBCCDD* that would be pure-breeding, and it should have fruit larger than either parent. Exactly the same possibilities would apply to flesh thickness or other quantitatively inherited characteristics. (Many agriculturally important characteristics are quantitatively inherited—most in fact. Fruit size, plant vigor, and yield are usually quantitatively inherited, for example.)

Of course, if the two parents I cross have the same genes for fruit size, doing the cross would not get me anywhere. So guessing which combination of parents would be most likely to contribute different genes for size instead of having the same genes for size is part of the skill in choosing crosses with which to start plant breeding projects. Usually, it's the more distantly related varieties that are more likely to have different genes for the characteristic of interest.

There was another reason why I thought that I might be able to get bigger fruits or thicker flesh just by intercrossing different delicata varieties and selecting. Fruit size and flesh thickness often seem to be generally related to vigor. If I merely managed to select a more vigorous variety it might have bigger fruits or thicker flesh just as a by-product of the extra vigor, even if no particular genetic differences for fruit size or flesh thickness per se were involved. It should be easy to select a more vigorous variety. Squash varieties are usually produced originally by methods involving a good bit of inbreeding, and inbreeding can lower overall vigor. If I outcrossed different delicata varieties, that could correct for the inbreeding. Then I would simply need to avoid inbreeding as much as possible in the rest of the project or correct for it by developing different lines independently and recombining them at the end. The varieties most likely to combine to make the most vigorous hybrids, and contribute the most vigor to varieties derived from their hybrids, are usually varieties that are more distant rather than close relatives.

Which varieties would be the best to use as parents in the starting cross? Well, of the delicata varieties available, I liked 'Sugar Loaf—Hessel' best, so that was the logical choice as one parent. The worst choice of variety to use as the other parent should be 'Honeyboat', which is very closely related. 'Sweet Dumpling' and 'Cornell Bush' looked like the best options. Nate France, whom I was working with at the time, had a line of 'Honeyboat' he liked, though, that we would be growing anyway just to eat. And in my experience theoretical knowledge only goes so far. There is much to be said for just trying stuff. So we decided to grow 'Sugar Loaf—Hessel', 'Honeyboat', 'Cornell Bush', and 'Sweet Dumpling' and cross them in all possible combinations. The characteristics of the F_1 generation of each cross would themselves tell us a certain amount and help us decide what cross or crosses to continue working with. Our 'Sweet Dumpling' seed turned out to be dead, however. So we grew and made all possible crosses among the other three.

We used an added trick. I had noticed that there was a lot of variability in flesh thickness on the fruits from plant to plant within each of the varieties. So as crosses to start our project we wanted crosses of particularly good individuals of each variety to particularly good individuals of other varieties. Nate made all those initial crosses, numbering each plant and recording the pollen source for each cross. At the end of the season we labeled each fruit as to the plant it came from. Then I took all the hand-pollinated fruits home as well as some samples of other fruits from each plant and evaluated each plant for fruit size, flesh thickness, flavor, and texture. The following year Nate and I grew about forty F_1 plants from the five crosses that represented the best individuals of each variety crossed with the best individual of one of the other varieties. Of all these plants, it was the crosses between the best individuals of 'Sugar Loaf—Hessel' and the best individuals of 'Honeyboat' that gave us F_1 hybrid plants with the most vigor and the biggest fruits. In addition, the hybrid fruits of some plants of those crosses also had substantially thicker flesh than I had ever seen on any delicata. So we chose those plants and crosses to work with further. (We had self-pollinated at least one fruit of every plant early in the flowering season, since we could not identify the plants that produced the best fruit until after the

fruit was harvested.) After we evaluated the fruits we saved seed from a self-pollinated fruit from each of the best plants. We planted that seed the next year to get F_2 plants from each of the best of our F_1 plants.

The F_2 plants showed a good bit of variability for fruit size as well as flesh thickness. We self-pollinated each of the F_2 plants, then after harvest again kept seed from those that produced the biggest fruits with thickest flesh. Then I intercrossed these two-generation inbred lines in the next generation so as to retain as much vigor from the outcrossing as possible (and correct somewhat for the two generations of inbreeding). Then we just mass-selected for fruit size and flesh thickness. That is, we just let the plants grow in a field with no other varieties of the same species nearby and allowed pollination to occur naturally. We evaluated each plant for fruit size and flesh thickness, flavor, and texture, and saved and pooled seeds from the best. Three generations later we had 'Candystick Dessert Delicata', a new pure-breeding open-pollinated variety.

Fruits of 'Candystick Dessert Delicata' are up to 3 pounds/1.36 kg. They are tan-and-green-striped. The flesh thickness is up to an inch (2.5 cm) thick. Because the fruits are about twice as big and the flesh is about twice as thick as for most delicata varieties, a 'Candystick' fruit has about five times or more edible food per fruit than any other delicata. I'm happy with half a squash or two halves for a serving depending on size instead of needing two or three entire squash for a serving as I had for preexisting delicata varieties. In addition, 'Candystick' plants are much more vigorous and productive than any other delicata variety. The fruits are as sweet as or sweeter than other delicatas. (I suspect from both flavor and cooking characteristics that 'Candystick' is the sweetest squash commercially available.) The flavor of 'Candystick' is distinctly different from the flavor of either parent and is spectacular. It's reminiscent of the flavor of Medjool dates.

Like other culinary varieties of the *C. pepo* species, 'Candystick Dessert Delicata' fruits need about a week or two to cure after picking. (If the fruits taste starchy instead of sweet, it's because they haven't been cured. Both the sweetness and the flavor of delicatas and other culinary *C. pepo* varieties require the curing period to develop.) And like other delicata varieties, 'Candystick' fruits hold their quality through December, but then deteriorate rapidly. That is, 'Candystick' stores well for a *C. pepo* variety, but doesn't have the extra-long storage life that can be had with the best-storing *C. maxima* varieties.

I wait until the vines die down before I harvest my squash. (However, make sure you harvest before freezing weather, as exposure to freezing weather shortens the storage life of squash.) I wipe the mud off each squash with a rag or paper towel as they come out of the field. As with other squash, if the fruits have wads of mud sticking to the spot where they touched the ground, rot often starts in that spot if the mud is not removed. I wipe off just the muddy spot, not the rest of the fruit. No other wiping or washing of the fruits is necessary. 'Candystick' squash have a heavily waxy surface that is good at resisting mold and rotting.

I store my Candysticks in mesh crates indoors. The fruits can be stacked up several layers deep within a crate as long as the crate is mesh of some sort that provides good air circulation, and as long as you turn or rearrange the fruits occasionally, especially the first couple of weeks after harvest. That first couple of weeks I turn a fan on

the stacked racks of fruits to help them dry out and cure.

To prepare 'Candystick' squash I cut the squash in half and quickly scoop out most of the seeds with a spoon, leaving the coarse debris lining the seed cavity intact to protect the flesh from drying out during cooking. Then I bake the halves upside down on the rack in a covered roasting pan. (I remove the debris around the seed cavity only after cooking when it scoops out easily.) The sugar content of 'Candystick' is so high it is easy to scorch the squash if you aren't careful. In addition, the final stage of the cooking should be at a lower temperature than for other squash because of the higher sugar content. Using a roasting pan with a lid (holes in the lid open), putting the squash on a rack in the pan instead of directly on the bottom of the pan, putting the squash cut-side down to retain moisture, and putting the roasting pan as high in the oven as possible (on a middle shelf, not lower) all help prevent scorching. I turn the oven temperature to 400°F/204°C initially for the first half hour, then turn it down to 325°F/163°C until the squash are soft. (The higher initial temperature cuts the overall cooking time, but will scorch the squash if they are actually allowed to come to that temperature.) I usually stack up as many cut halves as will fit in my roasting pan, since the squash are good both hot and cold. I can fit five squash cut lengthwise into halves in my pan. I stack them up, crisscrossing them so that the heat reaches all, keeping all away from the sides of the pan to prevent scorching.

I usually eat 'Candystick Dessert Delicata' squash plain without even salt or butter. There is no more need to add anything to a 'Candystick' than there is to a Medjool date. Both are so sweet and flavorful they are ideal by themselves. Cooked as I describe in a roasting pan so as to protect the squash from drying out or scorching, the texture of the squash is so fine-grained it is like a dense smooth pudding. I usually eat Candysticks as a dessert rather than along with the meal in the evening. A cold half of a Candystick with a chunk of cheese also makes for a great breakfast or lunch.

'Candystick Dessert Delicata' seed is available from my own company Fertile Valley Seeds, and from Adaptive Seeds and Southern Exposure Seed Exchange.

Another way I sometimes cook delicatas or other smaller squash, including 'Candystick', is to poke each fruit with a fork, then submerge the whole fruits in a big pot of boiling water until they are soft. I use a stainless-steel steamer basket turned upside down with a rock on top of it to hold the fruits beneath the water. This method doesn't give quite as spectacular a flavor as the roasting pan method but it is a good way to cook large amounts at once, as it saves the time of cutting. It's easier to cut cooked squash, and each person can cut their own at the table.

'Candystick Dessert Delicata' was derived from the cross that all my genetic knowledge and plant breeding experience told me was the least likely to be successful of the obvious possibilities. We tried it anyway. I still don't know for sure why that cross worked so well. In theory, it shouldn't have. The two parent varieties were too closely related to be able to contribute much in the way of different genes or added vigor to each other, I thought. Delightfully, the plants themselves thought differently.

Knowledge is more useful when you don't take it too seriously.

'Lofthouse Landrace Moschata' Squash

For my winter and spring gourmet squash supply I grow 'Sweet Meat—Oregon Homestead', a line of this Oregon heirloom I reselected for all its original virtues. (The general commercial strain of 'Sweet Meat' has now deteriorated and is no longer worth growing.) 'Sweet Meat—Oregon Homestead' is very vigorous. The fruits weigh up to 25 pounds/11.3 kg, have very thick flesh, a small seed cavity, huge (delicious) seeds and very sweet meat, and store through spring or longer, getting sweeter and more flavorful in storage. In addition, the seeds are able to germinate in cold mud and the plants thrive in cool or overcast weather in spring. I describe growing and using 'Sweet Meat—Oregon Homestead' in detail in *The Resilient Gardener*, where I also give my recipe for Perfect Pumpkin Pie made from these squash. 'Sweet Meat—Oregon Homestead' makes great pies, frozen squash, and fruit leather. And because of the thickness of the flesh it has several times as much food per squash as other squash of equal size. The seed is available through Fertile Valley Seeds, Nichols Garden Nursery, Adaptive Seeds, and Bountiful Gardens.

'Candystick' occupies the *C. pepo*/fall squash niche in our seed production field and on our tables. Since it cures quickly but doesn't hold as long as the best *C. maxima* types, we eat our Candysticks in fall and early winter. The *C. maxima* squash need a full month of curing before they are ready. So we start eating them only on and after Thanksgiving. 'Sweet Meat—Oregon Homestead' occupies my *C. maxima*/winter squash niche. Since these two species do not cross, we can grow big patches of both side by side in one field on one farm without having to do hand pollination to get pure seed. These squash are a major part of the home food supply for myself and for all the volunteers who help in the seed production field. One volunteer takes home a truckload of Sweet Meats each fall and eats them as well as turning the baked meat into fruit leather. The cooperating farmer has also begun selling 'Sweet Meat—Oregon Homestead' squash to the Corvallis school system, where the kids love them. ("They taste like they have honey and butter on them, even though we serve them just plain.")

The other main species of squash grown in our region is *Cucurbita moschata*, of which the most common types are various butternut varieties. Since these would not cross with our other two varieties, this meant that I could grow a seed crop of a moschata in the same field too, if I could find one I liked well enough. So for years I grew one or more different moschata varieties to try them out. But moschatas need more summer heat than pepos and maxes. We don't have enough summer heat to grow the bigger or sweeter gourmet types. All we can grow are the dinky or less flavorful types. Even 'Waltham Butternut', which is commonly grown here, is marginal for maturity, and the crop often fails to mature in a cooler summer. Until last year I never found a moschata I particularly liked.

Last year the *Cucurbita moschata* squash I chose to try was 'Lofthouse Landrace Moschata'. A New England seed customer suggested it to me. New England isn't much better for growing butternuts than western Oregon. But the 'Lofthouse' material produced big squash that actually matured nicely, he said. So I emailed the creator of this modern landrace for some seed. (A *landrace* is usually adapted to and grown in a specific region. It can be as uniform as most modern varieties, but in most cases is not. Landraces typically are uniform for some agronomical characteristics that matter, but are not uniform for many others.)

Joseph Lofthouse lives on a mountain in Utah. His growing season is very short. His summers are blazingly hot and windy. He tried every available *C. moschata* variety, but none could thrive in his horrible summers or mature completely in his short season. So he decided to breed his own. He got some of every *C. moschata* variety he could find, round types as well as classic butternut-shaped varieties, and planted them all together in a field where they could intercross to generate material with as much genetic variability as possible. Any genes for earliness that existed anywhere in any moschata variety would get incorporated into his project. Then he started selecting for earliness, and by mass-selecting rather than inbreeding to keep the genetic heterogeneity up and the material as vigorous as possible. (He had read my book *Breed Your Own Vegetable Varieties: The Gardener's and Farmer's Guide to Plant Breeding and Seed Saving*, so he knew all about these kinds of tactics.) Now, after some years of simply mass-selecting for earliness and for plants that can survive and thrive in his demanding conditions, he has 'Lofthouse Landrace Moschata'. What he wanted was a vigorous landrace squash that was early enough to produce in his short season, could stand his hot summers, and was highly productive. And that is what he developed. He did not select based on size or shape. He let the squash be any size or shape they felt like being.

But how would 'Lofthouse Landrace Moschata' do in Oregon? We don't exactly have blazing-hot summers. However, my New England customer had told me the Lofthouse material was the earliest moschata squash he had ever grown, and New England does have weather similar to mine in many respects. Time to just try the Lofthouse material. I was optimistic. We planted enough 'Lofthouse' in our seed production field to get a full seed crop of the squash, enough to sell, that is . . . if it grew and proved worth growing and selling.

'Lofthouse Landrace Moschata' *loves* western Oregon. The plants were long-vined and very vigorous. The fruits were various colors on the vines ranging from green or black to orange, but they all matured to the typical moschata buff color. The fruits were all different sizes and shapes. About three-quarters of the plants produced fruits that were butternut-shaped, that is, had a neck with solid meat and a bulb end with the seeds. The butternut-shaped fruits ranged from about 5 pounds/2.3 kg to about 15 pounds/6.8 kg. Some had short necks, some so short they were almost pear-shaped. Some had medium-length necks. Most had long necks, some long and narrow, others long and fat. The remaining quarter of the plants produced fruits that were round or oblate and that ranged in size from about 6 pounds/2.7 kg up to 30 pounds/13.6 kg. Most of these were ridged and looked like buff-colored Halloween pumpkins. A few looked more like buff watermelons.

'Lofthouse Landrace Moschata' plants are outrageously productive. Those that produced the biggest fruit usually produced eight or nine and matured them all. The plant that produced the 30-pound/13.6 kg fruits produced nine fully mature ones, for example—that is, 270 pounds/122 kg of fruit from a single plant. The plants that produced smaller fruits produced more of them.

Of about forty plants we grew, all plants but one matured their fruits. I have grown many *C. moschata* varieties, but this is the first time I have had any of the bigger moschatas mature here in a cool summer. And it had been a downright cold summer, even for maritime Oregon. I suspect that this moschata's genetic material may contain everything necessary to grow or adapt to conditions just about anywhere squash can be grown at all. I suggest it be tried in northern regions because of its proven ability to handle the cold springs we have in Oregon as well as our short growing seasons. I suggest it be tried in southern regions because moschatas in general are more resistant to the insects and diseases that trouble *C. pepo* and *C. maxima* varieties in those regions. And the genetic heterogeneity of the 'Lofthouse Landrace Moschata' means that at least some of the plants should carry resistance genes to most problems that might be relevant, providing starting material from which you can select according to your needs. The landrace was selected under hot summer conditions but has now proved vigorous in very cold summer conditions. And it's very early. It's theoretically possible to kill 'Lofthouse Landrace Moschata' plants, but you'd have to work at it. That is true resilience.

The squash required about a month of curing before they were ready to eat. Those that did not get eaten or processed into seed first stored nicely all winter. (How long they would last beyond early spring I don't know since we processed them all for seed at that point. Moschatas usually have a long storage life.)

When I cut into the first of the butternut-shaped squash, I was initially disappointed. It had even less sweetness than commercial store-bought butternuts, which are considerably less sweet than the best of the *C. pepo* and *C. maxima* varieties I usually grow. "Oh well," I thought. However, I was making a soup, and the butternut had so little sugar I could use a lot of it in the soup, just like carrots or potatoes. I don't usually use my supersweet squash for soups or stews. They are just too sweet for savory dishes. My soup that day ended up half squash, and it was great. The stew the next day ended up mostly 'Lofthouse' squash, and it was great too. And more 'Lofthouse' squash went into every soup and stew for weeks. And it was nice in salads. And many other ways. In fact, I found myself eating much more of it than I did the sweet squash and developing recipes for it. I ate about 200 pounds (91 kg) of 'Lofthouse Landrace Moschata' squash last winter, and never got tired of them. They had just the right amount of sugar and flavor to work well in savory dishes and combine with many other ingredients without overwhelming them.

I had thought I wanted a sweeter butternut with stronger flavor than what is grown commercially. In fact, what I really needed was a less sweet butternut with a more delicate flavor.

Most of the butternut-shaped squash from the 'Lofthouse' material have good texture, slight sweetness, a mild butternut flavor, and are great in the savory (non-sweet) recipes I present in the separate section on butternut squash cookery. A few of the pumpkin shapes also had good flesh quality, but most were coarse-fleshed and would make better ornamental pumpkins than edibles. For reasons I explain in the butternut cookery section below, I found the big butternuts with

Apologizing to a Squash

I've spent years sneering at squash varieties that are only slightly sweet and have only mild, delicate flavors. I hereby apologize to all such squash. I failed to understand them. Slightly sweet squash with mild flavors are amazingly versatile. The very sweet squash with powerful distinctive flavors are best for dishes that you want to be very sweet and need to have exactly that powerful flavor. Such squash make the best pies and sweet squash side dishes. However, I am coming to the conclusion that only slightly sweet squash with mild flavors may actually make the best staples.

Staples often are relatively bland. I think the mild flavors are easier to eat day after day without getting tired of them. White or whole wheat bread makes a more versatile staple than rye bread, for example. White-flour spaghetti and other pasta noodles are preferred by most to whole wheat versions. Blue potatoes are great occasionally, but they overwhelm most dishes and flavor combinations. The milder-flavored white or yellow potatoes are better day-in, day-out staples. So may it also be with the mild-flavored squash compared with sweet squash varieties.

long fat necks the most useful, and will focus on growing those in the future.

I sell 'Lofthouse Landrace Moschata' seed through Fertile Valley Seeds. To buy the seed directly from Joseph see his website garden.lofthouse.com.

Butternut Squash Cookery

One thing that defines the butternut cooking niche is how easy it is to cut the skin off the neck part of a butternut and chop it into pieces or grate it for using raw or tossing into a soup, stew, or stir-fry. Round-shaped squash don't really lend themselves to removing the skin and using raw. In addition, round squash often have ridges that further interfere with removing the skin when the squash is raw. Normally I don't toss cut sweet squash into fluid to boil because it dilutes the sugar and flavor, but that doesn't matter when the squash is only mildly sweet with mild flavor, and a mild, non-overwhelming flavor is what you want. The neck part of a butternut squash with its thin skin and no ridges is ideal for using raw.

There is an additional trick that can be used with butternut-shaped moschata squash. When you chop off a chunk of the neck, the remaining squash exudes a juice from veins around the edge that can dry into a protective resin. So when I use a butternut, I just chop off the amount of the neck I need for that meal. Then a few minutes later, after the remaining squash neck has exuded its juice, I spread the juice around over the cut surface with a clean finger. The juice dries into a resinous layer that seals the squash and protects it from drying out, rotting, or molding. The cut squash so sealed can keep for one to three weeks at room temperature. Many people want little butternuts because they only need squash for one or two people. But the most useful part of a butternut for removing the skin and using the flesh raw is the neck, which constitutes a much larger proportion of a big squash. The trick of sealing the neck of a big butternut means that these squash are just as suitable for a family of one or a few as small butter-nuts, and they are lots more efficient to produce, harvest, and store.

The recipes in this section were all developed for 'Lofthouse Landrace Moschata' butternuts. The recipes would also work well with many of the butternut varieties I've grown in the past that are also not very sweet or not very sweet when produced in my climate. They wouldn't work with moschatas like 'Tahitian Melon Squash' grown in the South, which are very sweet. In fact, they didn't even work very well with some of the commercial butternuts I bought just to try them on. Many commercial butternut varieties now are actually sweet enough so that they are no longer all that good for the soups, stews, and savory dishes that were the special role of butternut squash in the past. But these new butternut varieties are still not as sweet and flavorful and as suitable for pies or sweet side dishes as the best *C. pepo* and *C. maxima* varieties. The newer, sweeter moschata varieties are perfect for pretty much nothing.

Some traditional butternut recipes involve sweetening the squash to make pies or desserts. I see no point to using butternuts this way when there are so many other sweeter and more power-fully flavored squash that are more suitable for that purpose unless they are the only squash available. Instead, the recipes I present here take advantage of the fact that the butternut is mildly flavored, mildly sweet, and has a shape and skin type that lends itself to peeling the squash and using the pieces raw.

Enjoy.

Breakfast Butternut and Feta Cheese with Green Tea. One of my favorite breakfasts is just raw grated butternut squash with feta cheese crumbled and smashed in. I use about three times as much squash by weight as cheese. Use real feta cheese, which is always made from sheep's milk, and is imported. (American "feta" is usually made from cow's milk, which isn't the same thing at all.) Real sheep's-milk feta is rich and creamy and salty with a distinctive sour tang. Lately I've been using Pastures of Eden feta, which is imported from Israel and is economically priced at Trader Joe's. However, I've never met an Israeli or French sheep's-milk feta I didn't like. I eat this dish accompanied by plain green tea. The slightly sweet-sour tang and fattiness of the squash-cheese mix is complemented gloriously by the slightly bitter tea.

Butternut, Daikon, and Winter Greens Salad.

Grate and combine about equal volumes of the squash and daikon radish. Add a little lemon juice, olive oil, and salt. (For more crunch you can slice the squash and radish thin instead of grating them. Running the squash and daikon through a salad shooter works well, for example, and gives you more crunch.) You can use a mild fancy white vinegar such as wine vinegar, champagne vinegar, or white balsamic vinegar in place of or in combination with the lemon juice. You can also blend hazelnuts with a little water and use that for a homegrown or locally grown oil component. Eat straight or use as a base for whatever other winter vegetables you have. You can combine with finely chopped kale, for example, and/or overwintering leaf radish, scallions or shallots, or greens from overwintering onions or garlic. (You can also include "store-bought" vegetables such as red bell peppers or celery.)

To make a complete meal, I serve this salad along with some carbohydrates and some protein—a chunk of cornbread or a leftover potato for the carbohydrate, for example, and a chunk of leftover turkey or chicken or a piece of cheese or a hard-cooked egg or two as the protein. I also enjoy making a three-layer one-dish meal with the carb as the bottom layer, the salad as the middle layer, and bite-sized pieces of the protein course as the top layer. For the carb base layer I usually use cold cooked polenta or cold cooked chunks of white or yellow potatoes (baked or boiled). Cold cooked rice would also work. The polenta is what I usually use, as I grow and breed true flint corns I've selected specifically for spectacular flavor as cornbread and polenta. Use the polenta only if you have great polenta from a true flint variety. That means choosing the right corn variety and growing it yourself. (These corn varieties make great polentas: my own 'Cascade Ruby-Gold Flint' and 'Cascade Creamcap Flint' corns, and the heirloom true flint corns 'Roy's Calais' ['Abenaki Indian' flint], 'Rhode Island Whitecap', 'Narragansett Indian Flint', and 'Longfellow Flint'. Any store-bought polenta or cornmeal from dent, flour, or less pure flint types is so inferior as polenta it would only detract from the dish. See the corn chapter in *The Resilient Gardener*.)

Butternut, Daikon, and Fancy Olive Salad.

Slice or grate and combine equal amounts of squash and daikon radish (or amounts needed to balance the flavors). Then toss in generous amounts of fancy pitted olives such as Kalamata olives or green herbed olives. When you buy the olives, get some of the juice so you can use it as part of the dressing. Then mix in a little lemon juice, to taste. The olive juice and olives provide all the fat the salad needs to be delicious and filling.

Butternut and Sauerkraut (and Cheese). Grate the butternut and mix with sauerkraut (and cheese if you want it) in proportions to taste. Taste as you go. Add a little dry oregano. Or use fresh herbs to taste if you have them. Dress with a little olive oil if you don't use cheese, or add some other ingredient to provide fat, such as nuts, sunflower seeds, or hard-cooked eggs. Finely chopped kale or green onions or a shallot work well as additional ingredients. So does finely sliced red bell pepper. Or thawed frozen peas or beans or edible-pod peas.

I sometimes serve this salad on a bed of thawed, drained frozen (but still cold) eat-all greens. I sometimes buy a red bell pepper and make it into boat shapes, then fill them with the Butternut-Sauerkraut-Other-Stuff mix. I sometimes serve these kinds of salads on a bed of leftover cooked potato chunks or polenta to get a one-bowl meal with the exact amount of carbs I want. (Rice would also work.)

You can use any other salty fermented vegetables or (non-sweet) pickles you have instead of sauerkraut. If you are using commercial sauerkraut, I recommend Steinfeld's, which comes in jars. It has good flavor and is crunchy, not soft as are canned krauts. The crunch is part of what we are after.

Squashed Squash Soup. To make this soup I just microwave diced squash in a covered Pyrex bowl (no liquid added) until it is soft. Sometimes I use the squash as is with the whole chunks. Sometimes I run a potato masher or fork through part of the squash so that part is added to the broth of the soup and the rest remains in chunks. Then I add my homemade soup base or a can of commercial soup or stew. (My homemade soup base is usually meat and beans cooked without vegetables and seasoned heavily, enough so that it will be just right when diluted about two- or three-fold with freshly cooked vegetables. I make enough soup base for a few days' meals and combine it with the specific vegetables for each meal.)

Squashed Squash Soup with Greens. I start off cooking the squash in a Pyrex bowl as described in the previous recipe, then add my homemade soup or stew base. (Or a can of commercial soup or stew if I have no homemade handy.) I then microwave the dish a bit more to bring it to a boil and slice the greens if I'm using fresh greens. I add the sliced greens to the boiling soup or stew and microwave just enough longer to cook the greens. (About one to three minutes, depending on the greens.) In winter I sometimes use overwintering greens from the garden, but often use frozen eat-all greens. With frozen greens I add the thawed greens and the broth they were blanched in and frozen with to the hot soup as the last step, then microwave just enough longer to make the dish palatably hot, but not enough to start the greens cooking again. If I'm not going to eat the entire soup at once, I don't heat the main bowl after the greens are added. Instead, I dip out just one serving in a smaller bowl and heat that. I add a little lemon juice or vinegar to the individual serving just before eating.

Hamburgers and Condiment-Seasoned Squash.
I work some seasonings into the grass-fed burger meat with my hands. For 1 pound (453 g) of meat, for example, I might use 1 tablespoon chili powder, 1 tablespoon cumin, 1 teaspoon ground sage, and a little salt. I form the burgers loosely with indents in the center to correct for the fact that the edges lose more juice and fat than the middle and the burger will end up not being the same thickness throughout unless it starts thinner in the middle. Don't pack the meat more than necessary to get it to just barely hold together. Packing hard gives you a tough instead of a tender burger. You also want coarse-ground meat. Only a coarse-ground meat can give you a tender burger. (Some growers of grass-fed meat actually grind the burger meat so fine it is impossible to make a tender hamburger with it.) Also note: Good burgers are made from burger meat with substantial fat in it. Lean burger meat makes burgers that are both tough and flavorless.

I oil my frying pan with just enough oil to prevent the meat from sticking. I actually wipe all I can out with a paper towel so there is very little of this exogenous oil in the finished dish. Bring the frying pan up to medium-high heat and check it by tossing a few drops of water in it. The water drops should bounce and flash-vaporize. If they land and boil your skillet is too cold. If they vaporize without bouncing your skillet is too hot. If your fat smokes your skillet is too hot or you are using an oil with too low a smoke/burn point. (Corn, canola, soybean, peanut, and coconut oil all have high smoke points. Olive oil has a low smoke point and cannot be used for frying.) If you put the burgers into a skillet that isn't hot enough, they will not brown properly. And it is the browned meat that is the essence of the flavor of a great hamburger.

Place the burgers in the hot pan. While they are frying, dice 0.5 to 1 pound (227 to 453 g) of squash into half-inch (1.3 cm) cubes. (I like a monolayer of small cubes that fills the pan so that I have as much squash as possible, but it cooks in just a few minutes.) Flip the burgers only after the first side is thoroughly brown. Let the other side brown and the burgers finish cooking. Then remove the burgers. Add enough water to deglaze the pan and cover the squash cubes about halfway. Then cover the pan and cook the diced squash in the browned-burger-flavored fatty water, stirring regularly with a spatula while the squash is cooking. I usually add a teaspoon or so of cumin to the cooking diced squash. I start the squash cooking on medium-high heat and finish on medium-low. When the squash is soft, mash it with a fork. Add a little lemon juice or vinegar to balance the sugar in the squash. Then mix in a little ketchup to taste. Or mustard. Or pickle relish. Or Heinz 57 sauce. Or anything else you like on burgers. Or a mix of two or more of them. Then make a layer of the flavored mashed squash on each person's plate and put the hamburgers on top.

The idea is to get the mashed squash tasting slightly of the condiments so that a bite containing about half meat and half to two-thirds squash has just the right amount of condiments to complement the meat. Instead of a burger with just a small amount of vegetable powerfully flavored to make a condiment, you're eating the burger with a much larger amount of vegetable with less intense condiment flavors.

When I make hamburgers and eat them with just straight ketchup or other condiments, I usually end up eating a full pound of burgers. When I instead serve the hamburgers with the condiment-seasoned squash, I enjoy the dish lots more, and eat only about half as much meat or less, a good bit of squash, and about the same amount or less of the commercial condiments as I would have eaten if I had used them just straight on the meat.

Burgerdogs with Butternut Squash and Sauer-kraut. Why is ketchup on top of hot dogs or hamburgers so delicious? I think it is the combination of the sweetness, sourness, umami flavor, and complex-flavored fruit base of the ketchup with the savory umami flavor of the meat. However, I prefer to have my sugar tucked safely away inside the cells of a fruit or vegetable where my digestive system deals with it slowly in the fashion for which it has been designed. Ketchup is full of extra added (extra-cellular) sugar I would prefer to cut back on whenever possible. In addition, the ketchup is not actually as balanced as I would like. It seems both too sweet and too sour and too strong to me. I would prefer a sauce for my wieners that gives the same delicious effect as ketchup but is more dilute so I can eat more of it—so I can spoon up and gobble large enough amounts of the sauce so that it serves as the carbohydrate and vegetable portion of a one-pot meal instead of being just a condiment.

When you come right down to it, I don't really like the flavor of commercial wieners very much either, as proved by the fact that I won't eat them plain. Plain, they actually taste bad. Wieners are only good when complemented with ketchup or another appropriate condiment. In all truth, I eat wieners only for the ketchup. Once I realized that, various other possibilities become obvious. Why not put the ketchup on something I like better and, while I'm at it, something that might even be better for me? Can I use grass-fed local ground beef seasoned so as to be much more pleasing than any commercial wiener? Here's my answer—burgerdogs with butternut squash and sauerkraut. It's basically wieners and ketchup in which seasoned grass-fed ground beef is shaped like and substituted for the wieners, and sauerkraut and squash are substituted for the ketchup. (As the fact checkers at *The Atlantic Monthly* supposedly used to say, "There's nothing wrong with this article that can't be fixed by just replacing every word with some other word.")

The burgerdogs start as grass-fed hamburger I season and make into wiener shapes. I cook them on top of diced butternut squash in an 8-inch (20.3 cm) Pyrex bowl in a microwave oven. (See the end of the recipe for a non-microwave variant.) The burgerdogs release juice and fat that helps season the squash. Sauerkraut gets mixed in to provide the sour kick (and its own variant of umami flavor). The seasonings can be various. Today I used 1 teaspoon cumin, 1 teaspoon chili powder, and ¼ teaspoon salt for the 8 ounces (227 g) of grass-fed hamburger. I work the seasonings into the meat with my hands and shape the seasoned meat into three wiener shapes. An additional 1 teaspoon cumin goes on the 12 ounces (340 g) of diced squash. I arranged the squash around the edge of the Pyrex bowl, leaving a hole in the middle for more even cooking. The three burgerdogs go around the edge. (No water is added.)

I microwave the dish on high for four minutes, then turn the burgerdogs so whatever was above is below and whatever was toward the outside is inside. Then I zap on high for three more minutes. The meat and squash release enough juice so that the squash is now almost completely covered. Next I remove the burgerdogs temporarily and mash the squash in the juice. Then (after the cooking) I add sauerkraut or other fermented vegetables to taste until there is the right amount of sour zing in the squash-fermented vegetable combination. (When I use store-bought sauerkraut, I always get Steinfeld's.) I like about 8 ounces (227 g) of sauerkraut with 12 ounces (340 g) of squash.

Finally I slice the burgerdogs and mix the slices into the rest of the dish.

If you have a home in which no microwaves need apply, just cook the seasoned meat in chunks in a frying pan and remove the meat. Deglaze the pan with enough water to cook the diced squash. Then mash the squash and proceed as for the microwave version.

If you have leftover baked butternut squash or other slightly sweet squash you can use that instead of starting with raw squash.

You can also fry the meat as regular hamburgers and serve it on top of the squash-sauerkraut mix. You can also use sausage instead of seasoned hamburger. Don't use real wieners. They don't have enough of the right flavors to season the squash-kraut mix properly. Other seasonings I especially like mixed into the meat are paprika or sage.

Turkey Breast (or Chicken Breast) Butternut Cocktail. I love shrimp cocktail. But shrimp are expensive, and so is cocktail sauce if you buy it in little jars. You can make an excellent cocktail sauce with just Heinz ketchup, lemon juice, horseradish sauce, and a little black pepper. And I long ago discovered that I'm just as happy with cocktail sauce on pieces of white chicken or turkey as on shrimp. In fact, since I prefer the dark meat of poultry for eating straight or for soups and stews, this gives me something delicious to do with the white meat. In this recipe, I carry the substitutions a final step further by substituting mashed butternut squash for the ketchup in the cocktail sauce recipe.

I start off by zapping 8 ounces (227 g) of diced butternut squash in an 8-inch (20.3 cm) Pyrex bowl in the microwave oven for four minutes to cook it. I run a fork through it to mash it, then add about a cup of turkey or chicken broth. (My broth is from the chicken or turkey soup broth I made by boiling the meat in water with plenty of curry powder and sage and a little salt. My broth has delicious non-trivial amounts of fat in it. So if you substitute a commercial or wimpier broth you might want to add some olive oil or butter to the final dish.) I mix the broth, mashed squash, and shrimp-sized pieces of turkey together and zap the dish in the microwave one minute, stir, then zap again for forty-five seconds to heat everything up. Finally, after the cooking I add lemon juice and horseradish to taste. One lemon and a couple of tablespoons of horseradish sauce is about right. The idea is to have enough squash base so you can eat big spoonfuls of squash with the meat and the flavors are all balanced. The squash cocktail sauce becomes a major side dish, not just a condiment for the meat. Serve hot or cold. Makes great leftovers.

Planting by the Moon

I was once at a dinner at the home of Mushroom (Alan Kapuler) with a number of other avid gardeners. Mushroom was off in the kitchen cooking. Someone asked "Hey, what does everyone think about planting by the moon?" Everyone except me weighed in on the subject. Most were in favor of the idea of planting by the moon, some vociferously, though none seemed to actually be practicing it. And even those who wanted to believe in planting by the moon did not agree as to when you would actually end up planting anything. After the conversation ran down and there had been a short pause someone turned to me and asked me explicitly. "Carol, do you plant by the moon?"

"I plant by the sun," I said. "It's easier to plant when you can see the seeds."

After everyone finished laughing, I elaborated. In the Willamette Valley we can often plant the first planting of peas in February, for example. It requires watching for the break in the weather. There is usually only one such break of a few days in February. Miss it and your next opportunity might be delayed a month or two. I watch for weather that dries up the ground a little and is a little warmer, not for phases of the moon. If I had to have the phases of the moon right too, I would almost never be able to make that first planting.

In the Willamette Valley, nearly all the planting opportunities in spring are weather-limited. It rains often enough so that most of the time the soil is too wet to till or dig. When the soil dries enough so that it can be worked and planted, you better work it and plant, whatever the moon is or isn't doing.

After I expounded thus, Mushroom came back from the kitchen. So the question was put to him.

Talking to Your Plants

Some gardeners talk to their plants. I do it myself, though not from any special belief that the plants hear or understand. I just do it. For humans, talking is like grooming. Grooming, caring, expressing care, and nurturing are all mixed in there together. When we address a fellow human saying, "Hello. How are you? What about those Blazers? How's the family? Hot enough for you?" most of the time we aren't really asking for information. We are just engaging in a little casual verbal reciprocal grooming. Ducks do it more efficiently. They say "Wuk wuk." It means, "I'm here. You're here. We're all ducks, all members of this flock. We all belong here. I politely acknowledge your existence and our relatedness. We are all in it together. No danger is threatening. All is well."

It seems natural to me to talk to my plants while I nurture them. It makes me feel more connected to them, and I'm happier, whether they are or not. And perhaps they are. Perhaps when I'm happier in the garden I spend more time there watering and weeding my plants while I'm talking to them.

Did he plant by the moon? His response was as prosaic as mine.

"I plant when I have time," Mushroom said. There's a lot do in spring, he elaborated. There is soil preparation, putting out irrigation lines, planting and tending transplants in the greenhouse. When the weather is right and the soil is prepared and he has the time, he plants.

In our region I think the practical weather limitations are so great that even if planting by the moon actually helped some, it still would not be practical. We need to plant when weather permits, given that it doesn't most of the time. I think this is why even the people who would like to believe in planting by the moon here in the Willamette Valley by and large have never actually done so. Instead, they watch the weather longingly in spring, just like I do, and scurry out and dig or till, and plant on exactly the same days I do, whatever the moon is up to. There is no correlation between weather and phases of the moon. So planting when weather permits and is optimal means ignoring the moon. Clearly ignoring the moon does not prevent growing great gardens.

Do crops grow better ... even just a bit better ... if planted according to some phase in the moon, given equal weather (which would not normally happen)? We can't actually do a controlled experiment with two different phases of the moon at the same time to find out. Alternate experimental approaches are possible, but are difficult. All we can do easily is to note that many expert gardeners don't worry about the moon and have glorious gardens. And that, at least around here, those who support the idea of planting by the moon are not actually doing so.

Sometimes, however, agricultural beliefs work but for reasons different than the believer imagines. And often the beliefs work only in some regions but get passed along and applied outside their region of relevance. Let me imagine planting potatoes somewhere else that did not have such wet conditions in spring. Let me imagine a soil that dries up in spring and is suitable for planting potatoes most of the time for a couple of months in spring. In such a situation, it might be easy to procrastinate and put off planting the potatoes week by week until it is too late for an optimal crop. In that situation, if I believed that the potatoes should be planted on a particular magic day or at a particular phase of the moon, this might help ensure that the potato crop got planted in a timely fashion. So the false belief could lead to more successful potato crops, even if the potatoes themselves did not actually care at all about that magic date or the phases of the moon.

I usually don't assume plants need something without data showing that they do better with it. For me, the burden of proof is on the idea or assumption that is going to make gardening harder or more expensive. So in the absence of hard data, I assume the plants don't care about the moon. There is plenty of reason for some ocean creatures, especially shore or reef creatures, to care about tides, hence phases of the moon. On land what we usually see is plants caring not about phases of the moon but about moisture, temperature, weather, and sun. Many plants flower in response to a certain length of daylight, for example, thus ensuring that they flower at the right time of year. Some seeds require light of a certain spectrum in order to germinate, thus germinating only when they are near the surface rather than buried too deeply to emerge.

Non-knowing can be uncomfortable. I think we would often rather imagine we know, understand, and can influence something than admit that we neither understand nor can control it. So we develop beliefs and rituals that may be contrary to fact. Many people add unnecessary work that may sometimes be counterproductive, getting in the way of observation or looking for real solutions. There is much to be said for learning to be comfortable with non-knowing.

True Understanding

In ancient China, Gengsang was a disciple of Lao Tzu. When Gengsang was ready to go off on his own he built himself a hut on a hill overlooking a small village. He often sat peacefully on the hilltop and watched the activities of the villagers below as they tilled their fields and went about their business. He soon had several disciples of his own as well.

When fall came and the villagers harvested their crops, they left a generous portion of everything on a flat stone halfway up the hill. A disciple brought the offering to Gengsang.

"What's this?" Gengsang asked.

"The crops thrived, and the harvest is spectacular," the disciple said. "The villagers all say it is because of their sage who watches over them from the hill."

"I had nothing to do with it," said Gengsang. "The weather was good. There were no plagues of diseases or insects. These things sometimes happen and sometimes don't. I don't know any more about it than anyone else."

"Whatever," said the disciple. "Nevertheless, the villagers appreciate you. This could be a pretty good thing you've got going here."

"I don't think so," said Gengsang. That very day he picked up his few belongings and, leaving behind both villagers and disciples, moved off deep into the forest.

A few centuries later in ancient China a disciple asked Chuang Tzu: "Master, where did the universe come from? What is the purpose of life? Why is there pain and suffering? Where do we go after we die?"

Chuang Tzu responded: "The true master of life does not labor over life. The true master of fate does not berate fate. True Understanding uses Understanding to understand what can be understood with Understanding, and then stops."

*The skillful carver does little cutting. The skillful traveler leaves
few tracks and traces. The skillful binding takes few knots,
but does not come undone. The skillful speaker uses few words.*

道

CHAPTER TEN

Effortless Effort—The Eat-All Greens Garden

The No-Labor Garden—Just Sow and Harvest. The Nutritionally Most Important Home Garden Crop. Leaves Versus Heads or Stems. The Essential Role of Cooking. Using Greens in Soups and Stews. The Mess o' Greens. Harvesting and Handling Eat-All Greens. Freezing Eat-All Greens. Dried Greens and Herbal Teas. Lactofermenting Greens. Growing Eat-All Greens. Eleven Great Eat-All Greens Varieties.

Once upon a time in ancient China, Cook Ting butchered an ox for Prince Wen Hui. Ting made just a few swift movements and the ox fell into pieces.

"Amazing!" said the Prince. "I've never seen an ox butchered so swiftly or so gracefully. It was almost as if you were dancing instead of doing a piece of difficult labor! How in the world do you manage that?"

"When I first started butchering I saw the outside of the ox with my eyes," Cook Ting responded. "I had to hack and chop to separate the joints. It was hard work, and it took me just as long as everyone else. Then I learned about following the Tao.

"Now I see the essential nature of the carcass with my mind. I see just how the bones meet at the joints and are held together only by the tendons. My knife seeks and finds those spaces almost by itself. I slice gently in exactly the right places. The joints separate and the carcass falls apart almost spontaneously."

The No-Labor Garden— Just Sow and Harvest

One spring when I lived in town in Corvallis and had just a couple of small raised beds to garden on I ordered 2 cubic yards (1.5 cubic meters) of compost for use later in the season. I had it dumped on the concrete driveway, which I didn't use, having no vehicle in that era. I initially planned to cover the pile with a tarp to protect the nutrients from leaching out in the frequent rains. As I stood looking at the pile and the empty, unneeded driveway space, though, I realized that if I spread that compost into a layer about 6 inches deep (15.2 cm) I might be able to grow a crop on it before I needed the compost in a couple of months. Six inches of soil isn't very deep, but for a crop of something to be harvested quickly before the plants were very big, and at a time of year when regular rains and cool, moist air conditions would keep the layer moist, it might be adequate. So with five minutes of work I spread the compost out into a shallow bed on the driveway that gave me an extra 12 square yards (10 square meters) of growing space, at least for a couple of months.

It was mid-March, which in my region features variable weather. There are enough days with enough warmth to start some of the more hardy of the cool-weather-loving greens. However, many days are cold, and freezes can happen right up until about mid-May. What could I grow in a couple of months running from mid-March to mid-May? What might grow really happily and quickly at that time of year and produce a large amount of biomass for the space? I chose 'Green Wave' mustard. I love mustard greens, especially 'Green Wave'. And at that time in my life I just could not get enough mustard and other high-nutrient leafy cooking greens.

So I broadcast 'Green Wave' mustard seed over the entire bed. By that I mean I sprinkled the seed as uniformly as possible over the entire surface of the bed rather than making furrows or rows. To bury the seed very shallowly I gently bounced a rake over the surface of the bed so that I was disturbing the soil just a little, but not actually raking. (An ordinary leaf rake with springy metal fingers is what I used.) Then I let the seeds, the compost, and the rain do their work.

Usually in the past I had grown 'Green Wave' into fairly big plants and harvested individual

full-sized but young leaves for my soups, stews, or messes o' greens. The huge leaves are bright green, savoyed, and blazingly hot raw. However, when chopped and dropped into boiling water the leaves lose all their heat and have a delicious flavor. The stems of 'Green Wave' grown in this way are tough and unpalatable. As with kale, you have to pick the individual leaves. The broth left after the leaves cook in water is also delicious and makes good tea, soup, or stew. And there is real substance to 'Green Wave' leaves. A big bunch of spinach leaves cooks into just a few bites. An equal-sized bundle of 'Green Wave' mustard cooks down to much more food. In addition, the leaves of 'Green Wave' are several-fold bigger, often ten times or more larger than the biggest spinach leaves, so it is much easier to harvest serious amounts. 'Green Wave' also grows much faster than spinach. And in my opinion, as a cooked green, it tastes much better. The leaves of 'Green Wave' are too hot to use raw, however. But this has an advantage. The leaves are too hot for rabbits and deer. Only a creature who can cook can use 'Green Wave' mustard. So when you plant it you get the whole crop.

After about eight weeks my driveway raised bed had become a solid uniform stand of 'Green Wave' about 10 inches (25 cm) high. These plants were quite different from my usual crop of 'Green Wave', however. The entire top 7 inches (18 cm) of all the plants was soft, succulent, and edible, the main stem as well as all the leaves. I simply took a serrated kitchen knife and pan out to the garden and chopped off swaths of everything in the bed above 3 inches (7.6 cm) high. Then I took those bunches of greens inside, ran a knife through them to cut the greens and stems into

1-inch (2.5 cm) sections, and dropped them into my soup or stew during the last minute of cooking. Or I dropped them into boiling water, boiled them a minute, then drained them and dressed them with something delicious for a mess o' greens. (Meat drippings, salt, pepper, and vinegar, for example. Or any combination I might use on salads.) I did not weigh the harvest; however, from subsequent experience I think I got at least 30 pounds (13.6 kg) of edible greens from this 12-square-yard (10-square-meter) garden. This was actually the first time in my life I had enough greens to be satisfied. I ate a huge mess o' greens as the main course every day for a week, and froze enough to give me all the greens I wanted for much of the rest of the summer—all from a small shallow bed on top of part of a concrete driveway, and all in only two months.

That was my first eat-all greens crop. It had cost me no time or effort in weeding or watering. In fact I had done no work other than sowing the seeds and harvesting. And the harvesting and kitchen prep was also a small fraction of what was required for most greens. I cut off the top 70 percent of all the plants, and the main stems and all the attached leaves were all prime and edible. Because of the erect form of the 'Green Wave' variety, there was no mud or dirt on the harvested food, so I could just cut it up and use it without washing. The few small, yellow, older unpalatable leaves were down below the harvest line. I had happened to grow the plants in weed-free compost, but 'Green Wave' can perform equally well in any good garden soil. At the right time of year 'Green Wave' grows so vigorously it shades out any weeds.

Up until that first eat-all greens crop I had always accepted a fair amount of labor as the cost

of homegrown vegetables. Suddenly I had a new vision—I wanted gardens and garden crops that produced huge amounts of food on little or no labor. I wanted to just sow the seeds and go away, then come back at some point later and harvest. With the right varieties of greens, and the eat-all growing pattern, it turns out that this no-labor pattern is actually possible. During periods when there isn't enough rain, watering will be necessary. But weeding can be avoided, as can most of the labor usually associated with harvesting and kitchen prep. The individual eat-all crops are not only far more productive than the ordinary crops and styles of growing other greens, but in addition, you can plant a number of eat-all crops on one bed in a season.

For example, the right eat-all varieties can produce up to about 4.5 pounds/square yard (2.45 kg/square meter) of greens, with half to three-quarters of that amount being easy and normal. This means a single small bed with 12 square yards (10 square meters) of good garden soil in my climate can produce up to about 54 pounds (24.5 kg) of edible greens in eight weeks. And multiple crops per year are possible—about four in the Willamette Valley of Oregon, making for up to 216 pounds (98 kg) from the single small bed in the early-spring-to-late-fall growing season. Even if I lose about half the potential production through various imperfections, and I am too sloppy and disorganized to get four crops but instead get only three, that still comes out to 81 pounds (47 kg).

Furthermore, in many cases, the eat-all greens growing space doesn't actually cost any garden space. Their rapid growth and shade tolerance makes eat-all greens the ideal crop for interplanting or catch cropping, that is, using space that is available only part of the season until the main crop gets big enough to need it.

Here in the Willamette Valley I sometimes plant eat-all greens crops in mid-March on the land where I intend to plant tomatoes or squash later on. I harvest the greens in mid-May, then plant the warm-season crops. Eat-all crop beds combine naturally with big viney squash. Just put them in the middle between the rows. Harvest two months later and let the squash vines take over.

Eat-all greens should also be the ideal crops for growing in shallow containers on balconies or in rooftop gardens.

The greens growing style known as "cut and come again" shares some features with the eat-all method. In both methods the seed is broadcast in beds. And with both you clear-cut the entire top of the patch instead of picking individual leaves or pulling whole plants. With "cut and come again," however, the seed is broadcasted much more densely, and the plants are usually harvested when quite small—usually about 4 inches (10.2 cm) high. The total biomass produced by "cut and come again" is small, probably much less than 10 percent of what is produced by the eat-all method. The intent of the "cut and come again" method is usually to produce very young, tender prime salad greens. The "cut and come again" part refers to the fact that after the first harvest (with appropriate varieties), you may be able to get an entire second crop. Usually, however, the second crop has smaller, tougher leaves and more stems and isn't very good quality compared with the first crop. So many people harvest just one crop from each planting.

Microgreens are also broadcast into beds and harvested by clear-cutting the entire top of the patch. However, they are sown even more densely than for the "cut and come again" method, and produce even smaller crops of even younger greens for salads.

I believe the eat-all style of growing greens and the eat-all varieties will be a game-changer for all gardeners, but especially for those with small gardens. In addition, the high nutrient content of the eat-all greens crops as well as the very high yield makes them an ideal choice for growing in community or public gardens or in "food deserts" where greens are especially needed and are unavailable. And because of the very low amounts of labor for harvesting big batches at once and the minimal labor in the kitchen, the eat-all greens are also ideal for freezing. And it turns out that most also make good dried greens for use either in soups and stews or as herbal teas. Many who have very tiny gardens or short growing seasons may now be able to produce a year-round supply of greens for their families when they have never before been able to even imagine doing so. It should also be possible to produce a much larger variety of commercial frozen greens at more economical prices than the spinach or occasional package of turnip greens, the only frozen greens typically available.

I lucked out in many respects with that first eat-all crop. I happened to choose one of the few appropriate varieties and planted it at the perfect time of year for that variety to grow using an eat-all approach. And the spacing just happened to be perfect. I subsequently tried again and the mustard bolted on tiny plants because the time of year wasn't correct.

A good eat-all variety must germinate rapidly and grow quickly enough to outcompete and shade out weeds. (I usually grow in soil, not weed-free compost.) The plants must have succulent stems so that everything above the harvest line is edible. The plant form must be erect enough (or capable of growing erect enough when crowded) to keep the harvested part of the plant well above the soil so it is free of dirt or mud and can be handled in the kitchen without washing or sorting. And a good eat-all variety must produce a tremendous amount of biomass for the space and time. Furthermore, that biomass must be substantive, not mostly water. That is, the variety should not cook down to nothing (as does spinach). And time of year and spacing also matter with each crop.

It was a few years before I sorted out all the factors and could produce eat-all crops of even 'Green Wave' mustard reliably. Then it was the work of two more decades to test hundreds of different varieties of different greens crops to find more that could be used as eat-all types. In this chapter I present the results of this work to date—eleven good eat-all crops representing five plant families: several leaf-bred radish varieties, one mustard, two Chinese cabbage varieties, one kale, several gai lohns, two amaranths, huauzontle, quinoa, pea shoots, and shungiku.

I'll next outline why I think greens in general and cooking greens in particular should be the highest priority in home gardens. Then I'll discuss how to use cooking greens. Finally, having, I hope, convinced you how worthwhile and delightful it is to have plenty of cooking greens, I'll address the details of how to grow them using the eat-all greens garden approach.

The Nutritionally Most Important Home Garden Crop

If you have a bigger garden and need your garden to survive on, growing staple crops—crops with serious amounts of carbohydrates and protein such as corn, beans, squash, and potatoes—is essential. (Those are the crops I focus on in *The Resilient Gardener*.) If you have a smaller garden, however, you will probably buy or trade for most of your staples and just grow the crops that are the most important to you. Most important for many gardeners are tomatoes. However, what is most important nutritionally, I would argue, are greens, especially highly nutritious leafy cooking greens. Yellow vegetables are also valuable, but are not as critical. Good yellow vegetables are readily available commercially. In addition, greens also contain carotene. But greens contain other vitamins and nutrients that are found in significant quantities only in greens. And good cooking greens are not very available commercially, and tend to be expensive. A small bundle of a few leaves of kale in winter may cost $2, for example, which, by the time you consider just the edible portion, might work out to $20 per pound. I believe the most important thing most gardeners can do to enhance their family's health is to learn to grow and use lots of greens, especially cooking greens.

Most people these days don't actually know how to use cooking greens. Soups and stews are one way. You don't get nearly as much benefit as you could from your vegetable garden unless soups and stews made with homegrown vegetables are a regular part of your diet. However, a particularly glorious way to use cooking greens is the classic Southern "mess o' greens." If you add the mess o' greens to your repertoire and learn to dress them with all kinds of delicious things it can add to your eating pleasure and quality of life every bit as much as tomatoes.

Leaves Versus Heads or Stems

Green vegetables that are heads, that is, buds or flower buds, such as cabbage, Chinese cabbage (napa), and broccoli, are the major commercial greens because they lend themselves to handling, transporting, and storage. The loose-leaf greens such as mustard greens or kale generally have much higher vitamin content and are nutritionally more valuable than the head vegetables. (Lettuce is an exception. Loose-leaf lettuce is more nutritionally valuable than head lettuce, but neither is nutritionally very valuable compared with most other greens.) Vegetables that are more stem than leaves such as celery or bok choy are also generally much lower in nutritional value than greens that are mostly leaf rather than stems. The green leafy plant parts just have more vitamins in them than stems. And loose leaves generally have more vitamins than buds.

Home gardeners often grow cabbage, broccoli, and lettuce as their primary greens crops simply because they are familiar with them because of their being commercial crops. However, loose-leaf greens are generally easier to grow and usually yield much more for both the space and labor involved. Our home garden greens don't have to be handled much or transported farther than from the garden to the kitchen. And in my opinion most of the leafy greens taste better than cabbage, broccoli, or stem vegetables. Leafy greens also better lend themselves to preservation by fermenting, drying, or freezing. (Head vegetables are better for storing in root cellars, however.)

The Essential Role of Cooking

I like green salads; however, as best I can tell, I don't digest raw greens well enough to get much benefit from them. Admittedly, I discovered this back in the days when I had active celiac disease (that is, before I realized I had celiac disease and eliminated wheat and wheat relatives from my diet). Celiac disease damages the lining of the small intestine and causes poor absorption of many nutrients, so my situation was admittedly atypical. But perhaps my more extreme situation simply made a general human condition more obvious. Even though I no longer have active celiac disease, I still seem to benefit much more from cooked greens than from raw ones. That is, if I eat ample amounts of greens (far more than most people eat) but only in the form of salads, I soon develop cravings for cooked greens. If I have plenty of cooked greens I never develop cravings for raw greens. The greens I eat in salads simply don't seem to count nutritionally anywhere near as much as cooked greens, even when it is the same kinds of greens we are talking about. (Kale, for example, which I am happy to eat large amounts of, either cooked or raw.) The difference is obvious enough so that whenever I'm short of greens, I avoid "wasting" them by making salads. I cook them all. Even lettuce. When I'm short on greens I want the most nutritional value that I can get from those I have, and I'm convinced that, at least in my own personal case, this value comes from eating my greens cooked. (Note: My greens are usually cooked just a minute or two, not cooked to death. And I usually use the cooking water as well as the greens.)

Many cultures eat few if any raw vegetables, and for most of the rest, salads appear to be a recent introduction. Traditional Japanese and Chinese cuisine, for example, features a huge variety of vegetables, but they are cooked, stir-fried, added to rice or porridge, made into soups, or fermented, not eaten raw. Even cucumbers and lettuce are cooked. In Europe vegetables also seem to have largely been cooked until recently. The Romans and Greeks are exceptions. Romans were eating salads of leafy greens dressed with oil, vinegar, salt, and seasonings back in the days of the Roman Empire, and spread their salad eating and cultivation of salad crops throughout Europe. In Britain and most of the rest of Europe, though, the practice of eating raw greens generally seems to have not usually survived long beyond Roman defeats and withdrawal.

In post-Roman Anglo-Saxon Britain, for example, the staples were grain, pulses (peas and beans), and meat, with the wealthier eating more meat and the poor eating more grain and pulses. Onions, garlic, leeks, and cabbage were the main cultivated vegetables. The wealthy had multi-course meals featuring many meat dishes usually served with sauces containing cooked herbs and vegetables and even some whole courses of cooked vegetables. But nowhere on the menus that have survived from that era do we see any course that featured raw green vegetables.

The Anglo-Saxon peasant's meal was often a "pottage" (porridge or thick stew) based on ground grain and peas or beans and including or seasoned with the available garden vegetables, garden herbs, wild herbs, and wild greens. Wild brassicas such as mustard and kale were abun-dant and were especially valuable. Many of these wild herbs and greens were actually naturalized from domesticated versions grown during the earlier era of Roman occupation. (For the history of salads and vegetables in Britain I drew largely from *British Food: An Extraordinary Thousand Years of History* by Colin Spencer.) The Norman conquest, dispossession of peasants of access to common lands during the Land Enclosure era, and industrialization, which gave rise to a large class of urban poor with no access to gardens or common lands, all led to a general deterioration in the diet of ordinary people.

The British attitude toward raw greens up until fairly recently tended toward considering them inedible and unhealthful. Raw greens were for animals or savages. Civilized people did not eat them. I think it quite possible that raw greens actually often were unhealthful because they did not fit into the cultural pattern of fertilizing fields with raw manure, particularly with raw human waste, so-called night soil. Night soil was a major way of fertilizing vegetable plots in Japan and China. And vegetable farm production along the Thames downstream from London thrived because of the supply of night soil from the city. When I was a kid in Japan in 1958, we had myriad local vegetables, but we were explicitly warned not to use them raw because of the fact that they had been fertilized with night soil. The Japanese pattern of fertilizing with night soil was an excellent way to farm sustainably, and worked fine for a culture in which vegetables were all cooked. Salads don't fit into that pattern. It may be no accident that the eating of raw vegetables has spread first and most widely in the regions that have the least tradition of the use of night soil. In addition, salads have become generally popular mostly only after the introduction of and widespread use of fertilizers other than manure as well as after the germ theory of disease was understood and we had greater knowledge about how diseases are spread and took greater care with respect to using manure.

What do modern hunter-gatherers do with greens? Actually, there is very little that present-day hunter-gatherers eat raw. Certain fruits and certain fat-rich grubs are commonly eaten as gathered. But generally, greens are not eaten raw. They are collected and carried back to the hearth and added to the evening pottage or stew. Even the hunter-gatherer cultures reputed to eat meat raw generally turn out to eat only certain selected soft parts raw—liver and kidney, for example, the softest part of the animal carcass. Or the blubber of Arctic Ocean mammals, which is so soft it can be spread like butter. Dried or fermented meat is eaten without cooking. But drying denatures proteins, so is a kind of substitution for cooking, as is fermenting. Fish may be eaten raw, but fish flesh is soft, and it is often marinated in something acidic, which, like cooking, denatures, that is, somewhat predigests proteins. And usually only certain fish are eaten raw; fish is usually cooked. Modern hunter-gatherers the world over cook (or dry or ferment or acid-treat) almost everything, some fruits and nuts and the softest parts of game carcasses being the notable exceptions.

The earliest definitive evidence of hominid cooking dates back a million years, to the era of *Homo erectus*. It was recently discovered by Boston University archaeologist Paul Goldberg in a cave in Northern Cape Province, South Africa. (A good free article on the find is available on the Internet.

See "Archaeologists Find Earliest Evidence of Humans Cooking with Fire," by Kenneth Miller, *Discover Magazine*, December 17, 2013.) Many archaeological sites have established that we modern humans have been cooking throughout our entire two-hundred-thousand-year existence as a species. In his book *Catching Fire: How Cooking Made Us Human*, archaeologist Richard Wrangham presents his Cooking Hypothesis of human evolution in which he argues that control of fire and cooking was the central driving force behind human evolution. We began cooking, then evolved specifically in response to our diet of cooked food, he believes. And it was presumably the ancestors of *Homo erectus* who learned to cook and so evolved into *H. erectus*, a creature with the smaller gut (as evidenced by the smaller non-flared rib cage) and the small mouth, small teeth, small jaws, and weak jaw muscles characteristic of humans today. *H. erectus* also had a significantly bigger brain. Since both guts and brains are energetically costly, it might have been possible to evolve the bigger brains only after learning to cook and losing the bigger guts. Bigger brains might have led to yet more sophisticated cooking and food processing, leading to even more energy that could be spared on evolving bigger brains. According to Wrangham, the driving force behind human evolution was not walking erect or meat eating or cooperative hunting, but rather cooking and our evolutionary adaptation to a diet of cooked food.

Archaeologists all recognized that the skeletal changes in *H. erectus* must have evolved in response to a dramatically different diet that was easier to chew and digest. But up until Wrangham most thought it was more meat in the diet that was the improvement (perhaps as a result of cooperative hunting, for example). However, only the smaller gut fits with more meat being the important factor. Raw meat is tough. Predators and scavengers have large mouths, big sharp, pointed and ripping teeth, and big powerful jaws and jaw muscles. I think Wrangham's proposal that cooking was the big change makes more sense.

There is no particular reason to suppose that we evolved so as to be able to extract the maximum nutritional value out of greens when they are consumed raw. It would actually be astonishing if we did, given that we don't get maximum value out of grains, root crops, or meat when they are raw, and these are the foods where the subject has been better studied. There has been amazingly little work on the nutritional effects of cooking vegetables, and what there is tends to be simplistic. The values simply correct for the amount of water-soluble vitamins leached away in boiling, for example, which is irrelevant if you use the cooking water. Some nutrients are known to be released by cooking. Some are destroyed by cooking. Usually "destroyed by cooking" refers to canning or otherwise cooking the vegetables to death, not steaming, boiling, or stir-frying them briefly, and it assumes discarding the cooking water. Even such studies only describe the nutrients that are present before and after cooking, however. What matters is how well we absorb the nutrients. And there is little or no information on how well we absorb the nutrients from cooked versus uncooked vegetables. Still, we absorb nutrients better from starchy roots and tubers, grains, and meat when they are cooked, and by large margins, as do all other animals. (This is one of the reasons all animal feeds are cooked, even for animals that can digest the ingredients raw.

Another reason is virtually all animals from insects to rodents to chimpanzees prefer food cooked. We animals apparently have sensory mechanisms for detecting quality of food, and part of what we can detect are factors affecting digestibility. So we usually go for the cooked food, even if we are an insect or a mouse who has never encountered any cooked food before and did not evolve on it.)

It is clear that we don't actually need raw greens, as many people in cultures who never ate them have survived just fine for thousands of years. It's also clear that raw greens of appropriate kinds won't hurt us, at least as long as they are not laced with human feces, raw manure, or other sources of parasites and disease. I think that too few greens in the diet is one of the most significant problems with modern American diets today, but I admit that I am basically guessing from first principles. I also suspect, based on my own personal experiences, that lightly cooked greens (assuming the cooking water is used) provide us with more nutrients than uncooked greens, but there appears to be little or no research on the issue.

I encourage growing more greens. And I particularly encourage growing more cooking greens and making soups, stews, modern pottages, and messes o' greens a regular part of your repertoire.

Using Greens in Soups and Stews

It's really surprising how many gardeners don't make or eat soup. Soup can be amazingly fast as well as delicious. You can easily make a one-bowl meal for one or two people or a big pot for dozens. A bowl of soup can be an ideal breakfast or lunch or one-pot dinner. Or it can be one course of a dinner. (Just a great soup and a dessert can be a pretty good approach to dinner.) And soup allows you to use many vegetables that aren't quite prime enough to use for stand-alone dishes or preserving. Since I live alone at this point, I prepare a lot of soups and stews as easy one-pot meals.

Once you make soups a regular part of your diet you never have to waste or worry about leftovers again. Leftover dressed salad will be all wilted and unappetizing the next day . . . as salad. But just toss the whole thing into your soup during the last minute or two of cooking. Every salad, every leftover vegetable, can get a second life as a soup.

The boiling that happens in soup dilutes the flavors of many vegetables, allowing the use of vegetables that are too strong to use raw in more than small amounts. Fresh herbs, edible weeds, and roots or leaves that are a bit past prime or are strong-flavored can all make great soup. Carrot greens, cabbage greens, the greens of onions that you grew for the roots, sea kale, garlic scapes, horseradish leaves, and many wild greens are not really prime for salads, but are delicious in soup.

You can start a soup from scratch by cooking up a pot of dry beans or a chunk of meat. Or you can start with a can of commercial soup. I do both. I make my own soup bases by cooking dried beans or meat or both, then use the soup base as the base for different soups or stews for several days by combining the soup base with different vegetables. In between making my own soup bases I luxuriate in laziness and make soups by tossing lots of homegrown vegetables into a can of commercial soup. I usually end up with a soup or stew that is about half the commercial soup or stew and half homegrown vegetables. I discussed soups a bit already in the section on cooking with butternut squash in the Non-Knowing chapter; this time I'll focus on greens. I never make the

same soup twice, however, so my descriptions are approximate. Making a soup is a very ad lib process. It starts by looking to see what vegetables I have available. For me, the vegetables are the major component of the soup.

Commercial soups don't have enough vegetables in them. But that is perfect, because any vegetables in the canned soup are cooked to death during the canning, and it is especially impossible to can greens without cooking them to death. In addition, commercial canned soups are usually overly salty. If diluted with about an equal portion of real vegetables they can be great, however. If you are vegetarian you just use a vegetarian soup. I like meat, and often add extra meat to canned soups. The specific Progresso soups and the Dinty Moore stew I mention are tasty and have no gluten-containing ingredients. (Most commercial soups and stews have wheat as a thickener.) I describe each of three basic soups or stews starting off with a version made with a can of commercial soup or stew, then with versions using my own soup base.

I make a beef soup base by boiling a chunk of grass-fed chuck or some short ribs or a fatty roast (sometimes with marrow-containing soup bones in addition) in a pot of water (sometimes partly greens cooking water) with 2 or 3 tablespoons each of, for example, curry powder, chili powder, and cumin. (Many seasoning combinations are possible.) Often I add black-eyed peas or lentils too, since these go great with beef and do not require presoaking, and I prefer soup bases that are a mix of meat and pulses rather than just meat. When I include dry legumes I add the salt only after the legumes are soft. I chop the meat into strips 2 or 3 inches (5.1 to 7.6 cm) wide. Those pieces are small enough so that the seasonings reach more of the meat but big enough so that the meat stays succulent when it cooks. The grass-fed beef animal is at least twenty months old, so there is enough flavor for both the broth and the meat, at least if the meat has good amounts of fat. (Non-fatty meat does not give you a good-flavored soup. I have non-fatty cuts such as round steak ground and included in the burger. Commercial meat is usually from animals younger than twenty months, as is much grass-fed meat. It does not have enough flavor to taste great after being boiled. You also need a fat grass-fed animal, not a lean one. A better fatty acid profile is useless unless the meat actually contains fat. Lean grass-fed meat does not taste good and does not provide significant omega-3 fatty acids.) Usually about forty-five minutes is about the right amount of time to simmer the meat. Shorter and it is tough. Longer and too much of the fat and juice cook out and the meat is dry and flavorless. I test the meat at the forty-five-minute mark and stop the cooking by adding a frozen brick of the cooking water from greens.

For a turkey soup base I use about a quarter of a free-range turkey in the pot along with about 3 tablespoons of curry powder, water, and salt. (For full flavor the turkey should be four months old when butchered, not younger. Commercial birds are much younger and don't have enough flavor to flavor the broth very well let alone for the meat to be full of flavor after boiling.)

When the soup base is beans, I'm ordinarily using beans I've grown myself such as 'Gaucho' or 'Beefy Resilient'. I presoak them and cook them with the seasonings but with no salt. Salt is added after the beans have cooked. (They will not become tender if cooked with the salt.) Then the salt is added and the pot of beans is left alone

awhile to let the salt penetrate the beans. (See the bean chapter in *The Resilient Gardener* for more on dry bean cookery.)

I do not usually add carbohydrates to my soup bases. I cook the carbs separately and add just as much as I want to each individual day's soup. The carbohydrates are usually cooked potatoes, mashed potatoes, onions, cooked polenta, cornbread, or rice. Beans provide carbohydrates, but I seem to need more carbs than the beans provide by themselves. So I add some additional carbs to a bean soup.

When I use meat in the soup, I leave it in big chunks and make some meals with the meat as the main course. I may use some of the turkey chunks from the turkey soup to make turkey salad sandwiches (on white flint cornbread) using the same ingredients as for tuna salad sandwiches but with turkey instead of tuna. Or pieces of the beef or turkey might be heated up with spaghetti sauce and served over polenta or chunks of white cornbread or rice. When I want meat in the day's soup, I take a chunk of meat and cut it up into bite-sized portions only as it goes into the day's soup. Here are a few soups to give you an idea of the possibilities.

Oat and Greens Pottage. For the canned soup version I start with a can of Progresso Chicken and Wild Rice Soup. This soup tastes pretty good but has little chicken or even rice in it. I toss the can of soup in an 8-inch (20.3 cm) Pyrex bowl, add about two-thirds of the can's volume of extra water and ½ cup of gluten-free quick rolled oats. (Oats are naturally gluten-free, but commercial oats are usually cross-contaminated with wheat. Trader Joe's and bobsredmill.com both sell gluten-free rolled oats.) I toss in about

½ teaspoon of curry powder and some diced feta cheese for both the flavor and the extra fat and protein.

Next I wander out to the front yard and collect several handfuls of dandelion greens and 'Alexanders Greens'. It's late enough in spring that the dandelion leaves are thoroughly past prime, as evidenced by even the young leaves having central veins that exude bitter white latex when torn open. So instead of plucking whole dandelion leaves, I run my hand loosely up each dandelion leaf so as to strip off the leafy part and leave the central vein behind on the plant. The central vein is both more bitter and tougher than the leafy parts. (I never do the cooking in changes of water that some people use with dandelion greens. Instead I use the younger leaves, avoid the leaf veins when the leaf quality is marginal, and mix the greens with other types.) The 'Alexanders Greens' bolted a month ago, and those leaves are all past prime too. They would be too strong and tough for a salad. But they'll go great in the soup. Because both kinds of greens are a little older and tougher than they would be prime, I'll simply cook them a bit longer. In this situation, that means just tossing them into the pottage at the beginning instead of later, giving them four or five minutes of cooking instead of a minute or two. I want more protein in my soup and some extra flavor to cover the extra water and vegetables, so I cut up a chunk of feta cheese and add it to the soup. Oats bind lots of the seasonings and lessen their apparent intensity. When I use oats I usually add extra seasonings or flavorings.

I cover the dish and zap it in the microwave oven for about two minutes. Then I stir, and zap it again an additional minute. And so on for about four or five minutes or so of total cooking with intermittent stirring. Then I let the dish sit for a couple of minutes more before eating. (This treatment is aimed at cooking the oats without the thick stew spurting out along the way.) Then I add a little lemon juice or fancy white vinegar to bring out the flavors. (Something acidic added after cooking is the secret ingredient that turns merely good soups into delicious ones.) The time involved is less than ten minutes. My modern pottage ends up as a thick batch of delicious cooked rolled oats laced with greens and flavored with the commercial soup and the feta cheese. It's thick enough to serve on a plate if I want.

I usually have cooked polenta from my own homegrown polenta corns in the refrigerator, and more commonly use some of that rather than oats. If I have white- or red-corn cornbread around, I may toss chunks of that into the soup to provide the carbohydrate. (Yellow-corn cornbread doesn't go with the other ingredients as well.) Cooked leftover potatoes or rice also work. Or I might start the soup by dicing a potato and cooking it in the soup broth while I gather the greens. If I'm starting with cooked carbohydrate instead of dry oats I don't add the extra water, which was necessary only to hydrate the oats.

When I start with my own soup base I use the soup broth instead of the canned soup and include bits of meat as well as beans if there are beans. I don't add cheese. Otherwise, I proceed just as I do starting with canned soup. I end up with a thick oat pottage (or oat-bean pottage) laced with greens and bits of prime succulent grass-fed beef or free-range turkey and seasoned with my own delicious soup.

(Remember to add a little lemon juice or fancy vinegar after the cooking.)

Greens Broth Soup. This soup is lots of greens in a flavorful, mostly clear broth. Let's start with a version based on a can of soup, thawed frozen eat-all greens, and the thawed broth the greens cooked in. You can make the soup either in a pot on the stove or in a bowl in the microwave oven.

For the canned soup version I start with a can of Progresso Chicken and Rice Soup, add about a cup of thawed greens broth and some pepper, then add about a cup of thawed eat-all greens after the rest has come to a boil. After adding the greens I heat the dish just enough more to bring the soup up to a palatable temperature, but not enough to cook the greens any further. Any other dry or fresh seasonings you like can also go into the soup. As the last step I add a little lemon juice or fancy white vinegar to bring out the flavors.

If I'm starting with fresh raw eat-all or other greens instead of thawed, I bring the soup to a boil and add the greens, then bring back to a boil and cook the greens until they are done, which may take seconds to a few minutes depending on the type of greens and how prime they are.

Sometimes I cook some grass-fed burger into the soup or add bits of whatever cooked meat I have on hand. (Full-flavored grass-fed burger from an animal at least twenty months old when butchered tastes great boiled and releases juice and fat to make delicious soup.) Sometimes I add some sliced-up fancy olives to provide some extra fat and skip the protein. Most canned commercial soups don't have enough fat to taste good, and what fat there is isn't the kind I want. I usually remove the glob of fat in the canned soup and substitute my own. If I have no meat drippings or cheese handy I sometimes use KerryGold butter, which also goes great in soups.

When I start out with my own soup base I use just the broth of the soup and the greens with a little lemon juice or vinegar added after cooking. (Solubilized amino acids in the soup base provide the equivalent of all the protein the dish needs.) I like the soup to end up about one-third greens and the rest broth.

Cream of Greens Soup. For this I use a can of Progresso Clam Chowder, a can of albacore tuna with the juice, and enough greens so that the soup will end up about one-half greens by volume. I like to use a generous amount of curry powder in the soup. And some pepper. It's easier to use frozen thawed greens instead of fresh to make cream soup. I heat the clam chowder, tuna, and curry powder up in an 8-inch (20.3 cm) Pyrex bowl in the microwave oven, add the thawed drained greens, heat a little more, then add the lemon juice or fancy white vinegar. The commercial soup itself provides the carbs.

For a very thick cream soup you can just use fresh amaranth, huauzontle, or quinoa eat-all greens or wild lambsquarters. These leaves are so dry they absorb water when they cook.

If I have a turkey soup base handy, I use bits of the meat and just a little of the soup base broth in addition to the canned soup and skip the canned tuna.

Dinty Moore and Greens Stew. Dinty Moore is one of the few canned stews that is gluten-free. It has enough potatoes and carbs in it so I don't have to add extra carbs, and tastes pretty good. It can taste very good when properly jazzed up. So when I'm in lazy mode, I often use it as the base for a stew. I usually want more meat and fat in the stew as well as lots of greens. So I start off by removing the wad of fat in the can so as to substitute my own. Then I put the stew in an 8-inch (20.3 cm) Pyrex bowl along with some commercial salsa for complexity and heat, and put bits of grass-fed beef burger around the edge to cook as the stew is heated to provide both the extra meat and the good types of fat. (By the way . . . all the studies showing health disadvantages of red meat have to do with commercial grain-fed red meat overloaded with omega-6 fatty acids. Not grass-fed meat with a healthful balance of essential fatty acids.) After the stew is hot I fill the rest of the bowl to heaping with cut greens and stir, and have a bowl of greens dressed with stew. (I use far too many greens to fit them in the volume of the stew.) This will cook down to a dish that is about one-third greens by volume and two-thirds everything else. If the greens are one of the drier types that absorb water in cooking I add some extra water. If I want a soup instead of a stew I add extra water. If I have added no extra meat I add some KerryGold butter, or a little olive oil, or slice and add some fancy olives. As the final step, after the cooking I mix in the secret ingredient—a little red balsamic vinegar. (Red vinegars best complement a beef soup or stew. White vinegars best complement a chicken, turkey, or fish soup or stew.)

If I have some of my own soup base on hand, I sometimes still use the canned stew for its carbs, only add some broth and meat from my homemade soup.

Lentil and Greens Stew. Progresso Lentil Soup also has no gluten-containing ingredients. It's so thick that I use it the same as the Dinty Moore stew and end up with a lentil and greens stew. I usually also add some carbohydrate too. It's especially great with chunks of cold leftover potatoes reheated in the stew along with the greens.

If I have my own soup base available, I turn it into a soup or stew by just cooking up some home-grown vegetables in the soup base and adding some carbs. I might cook some diced potato in the broth, for example (or use leftover potatoes, polenta, or rice). Then I add lots of greens and cook them in the stew. I end up with a soup or stew that is about one-third greens and that has meat (and/or beans), broth, and some extra carbohydrate. Onions, carrots, radishes, or turnips can all provide a main ingredient. If you peel the carrots and cook them whole or just halved in the soup, then cut them after cooking, they will taste much better than if you slice them up first and then cook them.

The Mess o' Greens

A "mess o' greens," pronounced "messuhgreens," was what we called them when I was a high school kid in Georgia, and that is what I call them still. In print they seem to be usually referred to as a "mess o' greens," even though they are never pronounced that way, and I've honored that tradition. The greens are cooked by boiling, then dressed with any one of a number of delicious combinations. The high school cooks usually dressed them with bits of fried bacon, bacon fat, pepper, and vinegar, a southern classic. We kids loved them. We ate every bit we were served and would have eaten much more had they been available. Bacon and bacon fat are not today considered the epitome of healthful cooking. But there is much to be said for serving greens that your kids gobble down with relish, complaining only that there isn't more. A mess o' greens can be dressed in many ways, however, including with any salad dressing you like. Basically, the dressing usually contains some oil or fat, salt and pepper, seasonings of some kind, a little water or broth from cooking the greens, and something sour such as lemon juice or vinegar. That's flexible enough so that you can be as politically health-correct or -incorrect as you wish.

The greens in a mess o' greens can be any of the eat-all greens I describe later in the chapter. But they can also be any of the greens you pick by the individual leaf, such as turnip greens, kale, collards, beet greens or chard, edible weeds, wild greens, young leaves of perennials such as horseradish, sea kale, and lovage, fresh herbs, onion or garlic greens, spinach, or mixes of any of these. The mess o' greens is a spectacularly delicious way of eating large amounts of cooking greens.

Some people steam their greens, but I prefer to just boil them very briefly. Either will work for a small batch. For a huge batch, enough to freeze several pints, boiling is the way to go. You can't steam all that many greens at once. I usually fix a lot at once, and boil them in a canning kettle.

Basically I just bring a pot or kettle of water to a boil, toss the sliced greens in, bring the water back to a boil, then cook one to three minutes depending on the greens. (The water is not salted.) Not overcooking is important, especially with mustard greens, which become slimy if overcooked. Mustard and most kale greens take just a minute or so of simmering. Amaranth and the

Chenopodium greens (huauzontle and quinoa) take two to three minutes. 'Lacinato' kale (aka 'Dinosaur' or 'Tuscan') is an exception. It takes about eight to ten minutes to become tender. Some collard varieties also require the longer cooking. When dealing with an unfamiliar green, just start tasting some after one minute of cooking and keep cooking and tasting until the greens are delicious and tender. If they change from green to gray or brown or become too soft or even slimy you have overcooked them.

When the greens are cooked I place a strainer over a large bowl and dump the pot of cooking greens through the strainer to catch the greens and save the broth, the cooking water. After giving the greens a minute or so to drain I dump them onto a plate or into a shallow bowl and add ingredients to taste to make the desired dressing. I usually boil more than I need for one meal because I like them cold as well as hot, and because I freeze the excess.

How do I eat the greens? Myriad ways. Basically all the boiled eat-all greens are pretty mild after cooking. (The exception is shungiku.) So the key to using them is how they are dressed.

I often just dress a mess o' greens with a little olive oil, some Italian seasonings or dried oregano, lemon juice or a fancy vinegar, a little of the cooking broth, and salt and pepper. Sliced fancy olives can be tossed in. The dressing ordinarily contains oil or fat, something sour, salt (if there is no salty ingredient), and pepper (if there is no spicy ingredient).

One of my favorite dressings is crumbled feta cheese, oregano, pepper, and lemon juice. The feta cheese provides the salt, fat, and some of the sour tang.

The southern classic of bits of fried bacon, bacon fat, pepper, and vinegar was undoubtedly a classic because bacon was a staple; hard-smoked slab bacon can be stored without refrigeration. Bits of ham instead of bacon are also great.

Any meat drippings combined with salt, pepper, seasonings, and vinegar makes a good dressing for a mess o' greens.

Any salad dressing you like can be used. I make a Russian-style dressing with ketchup, balsamic vinegar, lemon juice, pepper, and mayonnaise or olive oil, for example.

The vinegar I use to dress greens is usually a fancy type such as red or white balsamic vinegar, red or white wine vinegar, sherry vinegar, or champagne vinegar. I often use lemon juice instead or a combination of lemon juice and vinegar.

You can use any fresh herbs you have instead of dried. Chop them fine and add them to the dressing. Likewise with onion or garlic greens. A finely sliced shallot goes great in many dressings.

I like to fry up some grass-fed beef into hamburgers, then deglaze the pan with water, add ketchup and a few other seasonings to the water, then mix the greens with part of the ketchupy fatty dressing, put the burgers on top, then put the rest of the dressing on top of the burgers. Served with a chunk of homegrown cornbread, it's a complete meal.

Sometimes I boil up some spicy sausages made from locally raised animals and add the greens during the last couple of minutes of cooking. Then I drain the greens, toss the sausages on top and cut them into slices so they release delicious fatty salty sausage juice onto the greens, and add a little vinegar. (The sausage provides all the salt and seasonings and fat.)

I like canned kippered herrings occasionally. I dress the greens with juice from the herring and vinegar, then put the herring on top of the greens.

Any meat course goes better on a bed of dressed greens. And with enough greens, I can be happy with just small bits of meat on top instead of wanting a major meat course.

Cheese sauces of various kinds can go on top of a mess o' greens.

Spaghetti sauce with or without meatballs goes really well on a mess o' greens. I like to sprinkle freshly grated Romano (sheep's-milk) cheese on top.

Hamburger gravy goes well with a mess o' greens. So does any other meat-containing gravy. Just mix the gravy and hot greens and serve over polenta, rice, mashed potatoes, microwave-reheated chunks of baked or boiled potatoes, or on top of chunks of cornbread.

With a layer of greens dressed with something that includes meat or cheese, the greens are a complete meal except for the carbohydrates. I usually eat my mess o' greens along with a chunk of cornbread or a side of homegrown polenta or with leftover cooked potatoes or with rice. If there is no protein in the dressing, sometimes a piece of meat or cheese or a hard-cooked egg or two also accompanies the greens. Sometimes I serve the dressed greens on top of polenta, rice, or potatoes.

Usually the mess o' greens, about three-quarters of a pound (340 g) of them (cooked) for a single serving for myself, is the main dish, with any meat, cheese, eggs, cornbread, polenta, rice, or potatoes being the side dishes.

I also like messes o' greens cold. I usually put some of the excess from each batch I cook into plastic freezer boxes, cover them with the cooled cooking water, then refrigerate a couple of the boxes for use the next two days and freeze the rest. The next day all day long I look forward to special things that can be done with the cold cooked greens.

I like a regular salad of some sort dressed with extra dressing and layered on top of a layer of cold unseasoned cooked greens. This is especially great when the salad is predominantly crunchy things like carrots, daikon radishes, celery, and apples. (Apples are a very versatile salad ingredient if they are the right kind, that is, one with a crunchy texture and a nice mix of sweet and tart flavors.)

The greens with canned kippered herring on them are also good cold.

I love tuna fish salad on top of a bed of cold eat-all greens. I add extra seasonings to the tuna fish salad to make up for the fact that the greens are unseasoned. (To make a great tuna salad I use albacore tuna, mayonnaise, lemon juice, vinegar, onion greens or a little mild onion such as Walla Walla or a shallot sliced fine, sliced apples, daikon radishes, celery, Italian seasonings, pepper, lemon juice, and a little fine white vinegar such as balsamic.)

Cold eat-all greens also go well in sandwiches. I make cornbread for sandwiches out of 'Cascade Creamcap Flint' corn, for example, using the Universal Skillet Bread recipe I give in the corn chapter of *The Resilient Gardener*. This recipe gives a cornbread that has no other grains and no artificial binders and holds together well enough to make sandwiches. (I prefer the mild neutral flavor of white cornbreads to yellow- or red-corn cornbreads for making sandwiches, as the flavor of white corn does not overwhelm the flavor of the sandwich ingredients. The flavor of yellow corn does not go well with most sandwich ingredients. If you can eat wheat, of course, you can just use ordinary wheat bread.) To use cooked cold eat-all greens in a sandwich I usually mix them with the condiments (ketchup, mustard, horseradish sauce, or pickle relish, or some combination of them) and use the condimenty greens as a layer in the sandwich along with the sliced cheese and/or meat. I love a layer of thin-sliced apples and/or daikon radishes in sandwiches too, along with the meat/cheese layer and the condiment-greens layer.

I like sauerkraut, though it's too salty and sour for my taste served straight. It's just perfect when served on top of or mixed with cold cooked greens.

One of my favorite things to do with cold cooked greens is to dress chunks of tomatoes with a Russian-style dressing (as described in the tomato chapter) and layer the dressed tomatoes over a layer of cooked cold greens.

Cold cooked or thawed frozen eat-all greens can also just be added to soups and stews. I add them after cooking the soup or stew, then heat just enough more to bring the dish back to palatably hot but not hot enough to cook the greens any further.

I usually save the cooking water left after boiling up a mess o' greens. I put some of it in a half-gallon (1.9 liter) glass jar and refrigerate it for use during the next couple of days. I dilute it a bit and heat it up instead of water and use it to make my tea in the morning. Or dilute it a little and use it straight as an herbal tea, or use it for the liquid in making herbal tea. Or dilute it a little and add fresh-squeezed lemon juice and sugar for a delightful green lemonade. Where the broth is particularly tasty (as for 'Green Wave' mustard, the chenopodiums, amaranths, and gai lohn) I freeze all excess cooking water in single-serving-sized containers to use in making soups and stews.

I use some of it in place of some of the water. I also add a frozen block of the greens water at the end of cooking a big pot of soup or stew so as to halt the cooking and prevent the ingredients from overcooking just from the heat retained in the pot after it is off the heat.

Harvesting and Handling Eat-All Greens

I use a serrated kitchen knife in the field to chop off the top of a patch of eat-all greens for harvest. The object is to chop at a level that leaves any yellowed or soiled leaves and tough stringy inedible stems behind and harvests only 100 percent edible plant tops with 100 percent succulent stems and all clean leaves. Typically, the plants are about 12 to 16 inches (30.5 to 41 cm) tall when harvested, and I chop them off at about 4 inches (10.2 cm) above the ground. I grab sections of plants with one hand, chop with the other, then lay the plants in a rectangular container with the stems all oriented in one direction. This greatly facilitates chopping the greens when I get them to the kitchen. (See the photo section, which shows the radish greens harvest.)

The harvested eat-all greens, like all harvested loose leaves, wilt rapidly if not deliberately handled so as to prevent wilting. I harvest last thing before leaving the garden. I use plastic containers that don't heat up in the sun as do metal bowls, for example. If the weather conditions are cool and moist, so much the better. I put a moist paper towel in with the greens to keep the relative humidity high and put the entire container into a plastic bag. When I get the greens home, I put the container inside the bag in the refrigerator promptly. The greens so handled and stored will usually hold in the refrigerator without deteriorating or wilting for two or three days. Without the moist paper towel the greens are unlikely to last even the rest of the day in the refrigerator without wilting.

Because of their sensitivity to wilting, eat-all greens will not usually be a good choice as a market crop. (When I envision commercial frozen eat-all greens crops I am envisioning a factory nearby that processes them within hours, as is usually done already for other frozen vegetable crops.) However, some of the eat-all varieties, such as the leaf radishes, can also be sold as market crops when the whole plants are pulled and sold as bunches. Gai lohn, having thicker stems, is often sold as a market crop.

The eat-all greens are much easier to prepare in the kitchen than most greens crops. They require no washing or sorting. Since the greens are all oriented with stems in one direction in the harvest containers, in the kitchen I just grab sections of the harvest, put them on a cutting board, and run a serrated kitchen knife through the greens perpendicular to the stems at about 1-inch (2.5 cm) intervals. If the plants are kept aligned in one direction during harvesting, slicing in the single direction is all that is required for preparation. This ease in kitchen prep labor in addition to the ease of growing and high yields is the reason why my frozen greens are usually all of eat-all varieties.

(Note: A heavy dose of a pollen you are allergic to on your greens can create problems if you don't wash the greens, perhaps even if you do. You can avoid the problem by timing your eat-all patches so that they are outside the pollination window of any nearby grass, trees, or other plants whose pollen you are allergic to.)

Freezing Eat-All Greens

Even if I have an ample supply of greens overwintering in the garden I like to have a good supply of frozen greens as well. Gardens can get temporarily buried under ice or snow. Overwintering greens are sometimes wiped out by hungry deer or rabbits. And my "home" garden is a short drive away from home; I don't go to it every day in winter. After heavy rains it's hard to even get into my garden without sinking up to my nostrils in mud. If I got in during those wetter periods I'm not entirely sure I'd ever make it out. And I can't go to my garden if the steep hill on the road below my house is icy, or if my van decides to develop problems. In addition, while I often enjoy cooking, sometimes I relish having food that is already all harvested and prepared. So I grow extra of the eat-all greens in summer and freeze substantial amounts of them for winter and for year-round fits of general laziness.

The blanching to prepare for freezing is identical to the cooking. I just dump any excess greens beyond what I need for the current meal onto a baking sheet and spread them out so they cool promptly. If the extra greens are left in a big pile those on the inside of the pile will overcook before they cool. I let the greens sit there until the cooking water has cooled. Then I pack greens into plastic freezer containers that are the right size for a one-meal portion. I use my hands to press the greens in as tightly as possible, then cover them with the cooled cooking water. I put aside a couple of boxes in the refrigerator for use as cold greens and freeze the rest. Where the greens are of types that produce the tastiest cooking water, I also freeze all the excess cooking water.

I use the thawed greens in any of the recipes for cold cooked greens. Or I heat them very briefly in the microwave oven for hot messes o' greens. Or I add them to soups and stews after the cooking, heating enough afterward only to warm the greens but not to cook them any further. Basically, you can use the thawed frozen greens any way you would use those freshly cooked.

Dried Greens and Herbal Teas

For my first attempts to dry eat-all greens I chopped them in 1-inch (2.5 cm) pieces just as I do for boiling. Then I piled them in layers on the trays in my Excalibur dehydrator. This dried the greens, but the stems as well as the center veins of all the leaves, even the small ones, dried into hard unpalatable little sticks that do not rehydrate easily. That is okay for using the dried greens as herbal teas, but not for eating them as a dried vegetable.

It's actually more practical to dry the greens on the dehydrator trays as whole harvested bunches of plant tops. It takes only part of a day for them to dry. When they are dry I then run my hand loosely down each plant, from stem end to top, in such a way that the chunks of dry leaves are stripped off and the stems and even the central veins of the leaves are left on the plant. What is harvested is a pile of dry leaf bits. Depending on the crop and variety, these dried bits can have more powerful and different flavors than the fresh or frozen leaves, and can greatly enhance the flavor of a winter soup or stew. The dry leaf chunks can also be tossed in a coffee grinder and turned into a powdered base for a soup or stew.

Some greens, such as 'Green Wave' mustard, huauzontle, quinoa, and amaranths, make delicious dry greens with distinctive flavors. Others, such as radish greens, give less interesting, blander dry

Lactofermenting Greens

Some of the eat-all greens varieties, the leaf-bred radishes, for example, are widely used for fermenting in Korea, Japan, and China, the countries that have bred them. I have yet to start experimenting with lactofermenting eat-all greens, but it is on my list. The eat-all greens should be ideal for fermenting because they are so easy to grow, so productive, and require so little work in the kitchen. I especially love the flavor of kimchi made with the traditional Chinese cabbage (napa). However, these head vegetables are fussier and not as vigorous as most vegetables I grow. One of the eat-all greens, however, 'Tokyo Bekana', is a loose-leaf Chinese cabbage. I'm particularly eager to try it for making kimchi. I think it should be possible to lactoferment any of the eat-all greens using standard recipes used with other greens.

greens. The edible weed lambsquarters, while not an eat-all, also makes good-flavored dry greens for soups, stews, or herbal teas using the same methods.

To reconstitute the dry greens I drop them into a soup or stew and cook them for a minute or two. They absorb a lot of water, so can be used to turn a soup into a stew or a thick dip. For example, I particularly like a dip made with a can of Progresso Clam Chowder, feta cheese, curry powder, and about 2 or 3 tablespoons of dried greens. Some garlic fits in pretty well too. I put them all in a Pyrex bowl and microwave until the cheese melts. The greens absorb enough water to turn the soup into a rich cheese dip . . . just perfect for dipping up with chunks of cornbread.

'Green Wave' mustard makes my favorite dry greens of those I have tried. Drying destroys the heat of the mustard, leaving a dry green that gives a rich succulent flavor when used either as an herbal tea or in soups and stews.

The dry greens rehydrate very rapidly when exposed to air. I pack them in small plastic bags that contain just enough for a few meals each.

I have no idea what the nutritional value of the dry greens is. Because of the extra labor involved in drying compared with either freezing or fermenting, I use the dry greens mostly as flavorings in soups and stews and dips, and for herbal teas. The 'Green Wave' herbal tea tastes a lot more delicious than most commercial herbal teas.

Growing Eat-All Greens

The plants of 'Green Wave' mustard in that first accidental eat-all crop I produced on a shallow temporary bed of compost on the driveway were quite different from the plants I usually grew. They were taller, the central stem was tender and succulent all the way to the ground, and all the leaves except just the very bottom leaf or two were still bright green and prime when the plants were quite large. This happens only when the soil fertility and water are not limiting but the plants are somewhat crowded. A row of plants will not grow into eat-all-style plants. The leaves spread out and many more of them droop over close enough to the ground so that they become muddy. The plants need to be just crowded enough so that they grow somewhat more erect and nearly all the leaves are held up far enough from the soil so they don't get dirty. The rapid growth is also essential so that all but the bottom leaf or two on each plant will still be succulent and prime when the plants are relatively large. To get a good eat-all crop, the plants have to be grown in a bed and crowded, but crowded just the right amount. If the plants are too crowded they compete too much with one another and fail to grow rapidly. Only rapid growth produces large plants that are almost entirely succulent and prime. I discuss spacing individually for each eat-all variety.

Most good garden soil is fertile enough to produce good eat-all crops. But they are also happy to grow in pure compost. My "garden" soil is often closer to field conditions than ordinary garden conditions because I breed corn, beans, and squash, and select, among other things, for varieties that are productive under regimens of little input. (I breed my corn, beans, and squash in part as survival crops. They are intended not just to produce the most delicious possible food right now, but also to help us survive the various natural disasters that are inevitable when considering longer time periods such as the next thousand years. The Northwest experiences an average of two or three Richter-scale-9 mega-earthquakes per thousand years, for example; one is actually due approximately right now, that is, within the next couple of hundred years. Such a quake would likely destroy all roads and bridges and leave us without access to much beyond our own neighborhoods for at least two or three years. My corns can produce well on just a tilled-in legume cover crop. We fertilize it only if the cover crop is inadequate.) I usually add some compost or organic chicken manure pellets to my eat-all beds before planting.

Ample water is the second key factor to growing a good eat-all crop, especially during the first week when the seeds are germinating. For a good eat-all crop you want the seed to germinate uniformly and the young plants to grow as rapidly as possible. You don't want the variability in germination and growth you get when the plants are fighting for water. For spring-planted crops in the maritime Northwest there are usually ample rains to provide water naturally. The problem is more one of finding a dry spell long enough to dry out the soil enough for turning over and planting. I have in mind constructing a movable low plastic tunnel to put over beds I want to work and plant so that I can dry out beds and plant them with or without cooperating weather. I just haven't gotten around to it.

When I plant eat-all greens crops in summer, our period of no rainfall at all, I hand-water them daily the first few days to keep the seed moist enough to germinate. (I use a hose with an attachment that can be set for a fine gentle spray.) Seed of the eat-all greens varieties ranges from very tiny (such as amaranth) to still pretty small (such as radish). The tinier the seed the more important it is for the seed to be just barely covered with soil. Just barely covered seed dries out readily. Once the plants have germinated I hand-water gently every other day for a week or so, then twice per week. All this is only for the summer-planted eat-alls. The spring-planted ones usually get enough water naturally. Even under dry conditions, with the watering needed, the eat-alls are a water-conserving crop, however. All the water is put on just a small concentrated space that produces a huge yield rather than spread over a much larger area that produces smaller yields per unit of land or amount of total water.

When I sow corn, beans, or squash, I operate in such a way so as to not have to bend over or kneel or get down on hands and knees. Not so for the eat-alls. Sowing them uniformly is critical to getting a good stand. But the total space is small. So I bend over, or get down on hands and knees and sprinkle the seeds with my fingers as uniformly as possible.

To bury the seed appropriately, I bounce a rake over the bed at about 1-inch (2.5 cm) intervals. I'm talking about the kind of rake with springy metal fingers that is used for raking leaves. I use it to disturb the soil only. I don't draw it at all or rake with it. If you do that it buries the seeds too deep. I usually sow at variable density, with one edge of the patch sown at what I think is likely to be too dense and one edge at what I think is likely to be too sparse, with increasing density from one side to the other. Usually much or most of the patch will be close enough to perfectly spaced. If some of it is too dense I just hoe it down and resow.

Usually when the seeds germinate there will be occasional clumps or areas where there are too many plants. I bounce the rake over these areas a few times to kill most of the plants, thus thinning the rest.

Polycultures of different eat-all varieties don't generally work as well as well as monocultures of one variety. Uniform stands and uniform growth rate are the keys to producing good eat-all crops. Anything sown along with leaf radishes, for example, just gets overgrown by the radishes. Each eat-all variety is prime at a different size and bolts at a different time. So I grow my eat-all patches as single-variety monocultures. However, I often put several blocks of different eat-all varieties in a single bed. And eat-all varieties are great for sowing in between squash rows. They can grow a crop and be harvested before the squash need the space later in the season.

Timing also matters. Amaranth grows well only after the weather has warmed up for good in the late spring. It can be planted all summer as long as there is warm weather. A few eat-all crops can tolerate even hard freezes and can be planted in early spring: pea shoots, leaf radishes, 'Green Wave' mustard. Many of the brassicas will bolt on small plants if planted just before the summer solstice but can be planted well before then, or after. I give details in the sections on the specific varieties.

There are three different patterns I use for harvesting eat-all greens. The simplest, already described, is to cut down the entire patch or large swaths of the entire patch. Then I use a shovel or my peasant hoe to hoe down or turn under the debris and replant the entire patch with another eat-all crop.

An alternative is to just leave the plants after the harvest, give them a good watering, and let them sprout a second crop of leaves. This pattern is referred to as "cut and come again." I usually don't do this because the leaves are not as prime and the yield per space, time, and labor isn't as good as I get from just replanting with a new eat-all crop. The harvest on the "come again" crop is more of individual leaves that have to be picked rather than an eat-all crop that is basically mowed. And the leaves are usually smaller and less prime than eat-all greens because the plants at that point are bigger and overcrowded.

Rather than harvesting an entire patch, though, I often harvest in such a way that I thin the plants appropriately and leave the remaining, wider-spaced plants to produce an additional harvest of greens or a root or grain crop. For example, given extra spacing, 'Shunkyo' dual-purpose radish can produce a nice crop of red radishes that can be sliced and tossed in the pot with the greens or eaten raw. For amaranth or quinoa I sometimes harvest-thin to appropriate spacing so that the remaining plants go on to produce a grain crop. With huauzontle, I sometimes harvest-thin to appropriate spacing, top the remaining plants, then let them produce tender green shoots and leaves on big sturdy bushes for the rest of the season. (Details are given in the following section for each individual eat-all variety.)

Eleven Great Eat-All Greens Varieties

Leaf Radishes (*Raphanus sativus*). I've become a big fan of leaf radishes, radishes bred specifically for the greens. Leaves of leaf-bred radish varieties have more and bigger leaves than root-bred varieties, and the leaves are more upright, more succulent, and less prickly. Leaf-bred radish leaves are widely used in China, Japan, and Korea for stir-fries, soups, fermented greens, and kimchi. They're also good in salads and as micro and baby-leaf greens. The young leaves have a distinctly radishy flavor that is great in salads or sandwiches. Picked at the bigger eat-all stage and cooked, the flavor is mild and the greens can be used in any of the ways I described for cooked greens.

Leaf radishes germinate and grow much faster and more vigorously than any other greens I have grown. Furthermore, they are hardy. They can be planted from as early as the ground can be worked in spring through fall. (Exception: They may bolt prematurely if planted shortly before the summer solstice.) Here in the Willamette Valley of Oregon, mature plants with big radishes will stand through the lighter freezes in early winter and only die out after

hard freezes in midwinter. Younger fall-planted radish plants with no radish roots will stand all winter and will continue growing whenever the temperature is above freezing.

The leaf radishes available now are all hybrids. I am in the process of dehybridizing them. Meanwhile, I recommend growing 'Shunkyo Semi-Long', an open-pollinated dual-purpose Chinese heirloom variety. (It's available from Fertile Valley Seeds, Johnny's Selected Seeds, and Wild Garden Seeds.) 'Shunkyo' makes tasty 4-inch (10.2 cm) long 1-inch (2.5 cm) wide red roots in about forty-five days. The root is sweet as well as pungent and is good raw or cooked. I like to slice the roots and cook them along with the greens. Generally, though, to get the roots takes spacing a little wider than is optimal for eat-all greens. So I sometimes harvest-thin an eat-all 'Shunkyo' patch in such a way as to leave some plants a little longer and with a little wider spacing to give me both greens and roots.

To grow 'Shunkyo' as an eat-all crop, I broadcast the seed in beds, thin to about 3 inches (7.6 cm) apart in all directions, and harvest the top 8 inches (20.3 cm) of the entire bed when plants are about 12 inches (30.5 cm) high, at roughly six to eight weeks.

There are also four hybrid leaf radish varieties available in the United States—'Saisai', 'Four Seasons', 'Hattorikun', and 'Pearl Leaf'. Of these, the one that is most readily available is 'Saisai', which is sold by Fedco Seeds. The other three are sold by Evergreen Seeds, a specialist in Asian vegetables. 'Saisai' and 'Hattorikun' are also sold by Kitazawa Seeds, another specialist in Asian vegetables. The variety names and catalog pictures suggest big differences in leaf appearance and seasonal growing niches, but in my garden these varieties all grow and look pretty much identical. All can be planted early spring through fall, and all will overwinter in my garden from fall plantings. All of these varieties will produce a big white daikon radish if planted in early summer and left in the ground long enough. But the radish is a winter radish, and doesn't develop in anywhere close to the same time frame as the eat-all greens crop. If you want radishes as well as greens, plant 'Shunkyo'. It's open-pollinated, and its radishes develop promptly and are also much tastier.

'Indian Spinach—Red Aztec Huauzontle' (*Chenopodium berlandiera*). This is a relative of lambsquarters (*C. alba*), but has bigger leaves, more tender stalks, and bolts much more slowly, so has plants that stay succulent and prime longer for use as greens. It's a native American heirloom from Mexico. The word *huauzontle* (pronounced "wuh-zont-lay") refers to a traditional use of the flower buds for stir-fries, but I don't find that use compelling. I suspect that this line has changed with respect to that characteristic, and/or development of really big buds needs a more southerly latitude. Instead, I use the variety as an eat-all greens crop. It is also good in salads. And it makes especially delicious dry greens and herbal tea. The young seedlings are red, but when plants are much beyond the seedling stage the leaves are green with a reddish tinge, not red.

'Red Aztec' can be planted from late spring to early fall. I broadcast in beds, thin to about 4 inches (10.2 cm) apart, and grow to about 12 inches (30.5 cm) high, then harvest the entire top 8 inches (20.3 cm), which is all succulent stalk and leaves. It takes about eight weeks for an eat-all crop under my coolish summer conditions. After harvest, you can pull or turn under all the plant stumps or debris and replant the bed. However, I prefer to harvest-thin so as to leave a few plants spaced at about 12 inches (30.5 cm) apart. I top these. They grow into nice bushes that produce tender shoots for greens all the rest of the season.

Quinoa (*Chenopodium quinoa*). This is a native American grain crop that likes cool summers with cold nights. It's a relative of 'Red Aztec Huauzontle' (*C. berlandiera*) and the edible weed lambsquarters (*C. alba*). The plants and/ or flower heads of quinoa can be green, bronze, gold, or red depending on varieties. For producing a grain crop the variety must be chosen for the specific area. 'Temuco' and 'Faro' are varieties that do well in the Willamette Valley, as do two of the varieties bred by Frank Morton, 'Red Head' and 'Ore de Valle'.

Quinoa seedlings are fairly frost-hardy and don't need as much heat to grow as does their relative 'Red Aztec Huauzontle', so they can be planted earlier in spring. However, quinoa bolts pretty promptly. To use it as greens I sow and grow it just like 'Red Aztec', but in early to midspring only (March or April). And instead of clear-cut harvesting, I usually harvest-thin to create rows with the remaining plants 1 foot (30.5 cm) apart in the rows and let the bed go on to produce grain. The greens taste and are used just like those of 'Red Aztec Huauzontle'. However, because the plant bolts so quickly, I sow it as an eat-all greens crop only in spring, and then only if I also want the grain. 'Red Aztec Huauzontle', which is very slow to bolt, is more versatile as a pure eat-all greens crop and is considerably more widely adapted than is quinoa. However, if you are growing quinoa for grain anyway and you initially sow and tend it eat-all style, you can get a good greens crop as well as your grain crop from one planting.

'Green Wave' Mustard (*Brassica juncea*). This is the best mustard for eat-all growing. No other mustard grows as fast, bolts as slowly, or has the necessary erect plant form. 'Green Wave' has large, broad, savoyed bright green leaves and cheerfully sneers at and outgrows weeds. It is fiery hot raw, but mild and delicious when grown eat-all style and cooked briefly. The heat is also destroyed when the greens are dried. It's my favorite dried green, my favorite dried for herbal teas, and when the greens are cooked, the cooking water broth is bright green and tasty.

In the Willamette Valley, 'Green Wave' should be planted mid-March for an early-spring eat-all crop. If it is planted April or later but before the summer solstice the plants will bolt before making a good eat-all crop. However, it can be planted again after the summer solstice through fall in mild-summer areas. In hotter summer areas it might be best planted as a fall crop.

I grow it with the same sort of spacing as described for leaf radishes. Fall-planted beds can last through the mild freezes of early winter.

Small amounts of young leaves can be chopped fine to use in salads or salad dressing, in which case you can skip the pepper.

'Green Wave' is more vulnerable to over-cooking than the other eat-all greens. When overcooked it turns brown and becomes slimy. I dump it in the boiling water, just barely bring the water back to boiling, then call it cooked.

'Green Wave' became my most important greens staple during an era in which I gardened in an area overrun with deer. The deer ate all the greens including other mustards as well as tomato and squash plants—everything except garlic greens and 'Green Wave' mustard. My duck flock ate all those things too, if given a chance, and even ate the garlic greens into the ground. But even they never tasted the 'Green Wave' more than once. The human–'Green Wave' relationship depends completely on the fact that we are a creature that cooks.

Pea Shoots (*Pisum sativa*). I use the variety 'Oregon Giant Sugar' because it grows rapidly and has huge leaves and is the main variety I use for growing edible pods. However, any variety that does well in your area that has big leaves and that is short to medium in vine size will do. (Varieties with small leaves or tendrils instead of leaves or that are tall types with long stems between each node of leaves don't produce as much biomass as short- or medium-vine varieties with big leaves.) I've used 'Austrian Field' pea, which is usually a winter cover crop, as a fall-planted eat-all.

Here in the Willamette Valley I can sow 'Oregon Giant Sugar' as an eat-all crop from early spring through fall. I usually sow it as an eat-all first thing in spring primarily, though, when there are not so many varieties that are appropriate. It's easier to get good eat-all crops of pea shoots in early spring than summer, as warm weather can toughen the stems. In areas with hotter summers, pea shoot eat-alls will likely be a spring crop only.

To use as an eat-all pea shoot crop, I sow in a wide row or bed at about 2 inches (5.1 cm) apart in all directions. When plants are about 6 inches (15.2 cm) high, I harvest the top 4 inches (10.2 cm). The yield of biomass is not nearly as large as with most of the other eat-all varieties. However, the flavor is quite different. The greens are good raw in salads, and have a flavor similar to the raw pods only without the sweetness. Boiled briefly, the greens taste pretty much the same as the pods do if boiled briefly. Boil them longer and the greens take on the flavor you get in canned peas. I like all those flavors. I use the greens in salads, soups, and stews, not usually for messes o' greens.

I've tried harvest thinning an eat-all pea patch so as to grow a crop of pods from the same planting, but it doesn't work well at all. The peas just tangle together too much to allow cutting in a controlled way. So I plant my pea patches that are intended for pea production as separate patches.

(If you use 'Austrian Field' peas as a winter cover crop you can pick the top few inches of the plants for use as greens throughout winter and spring.)

'Tokyo Bekana' (*Brassica rapa*). This variety is a very vigorous, fast-growing, unfussy loose-leaf Chinese cabbage that is mostly leaf instead of stem. It has a distinctive yellow-green color, great flavor, and crunchy texture. It takes about thirty days to the eat-all stage; forty-five days to loose heads. I think this variety is the best Chinese cabbage for salads. It's also great in stir-fries, soups, and stews, and should be excellent for kimchi. I'm guessing that 'Tokyo Bekana' is more nutritious than most Chinese cabbage since it is more leaf and less stem. It bolts fairly quickly. Flower scapes are also edible. The color

of 'Tokyo Bekana' is so beautiful I find myself sometimes using it as a catch crop just to add more of its glorious splashes of bright chartreuse to the landscape. I plant it late spring through early fall. Instead of clear-cut harvesting it, I thin-harvest and leave the remaining plants to grow into bigger plants I harvest as whole loose-leaf Chinese cabbage plants.

'Spring Raab' (*Brassica rapa*). I've tried this as a midspring-planted eat-all green, and it does work. However, it does not grow as fast, look as beautiful, or taste as good raw as its same-species relative 'Tokyo Bekana'. Nor does it yield as well as most of the other eat-all varieties. It also bolts much faster than most of the eat-all varieties. In short, 'Spring Raab' works as an eat-all, but there are lots of other varieties that work better.

Gai Lohn (*Brassica oleracea*). Most *B. oleracea* varieties have central stems that are too tough to work as eat-all greens crops. Gai lohn, however, also called kai lohn, Chinese kale, and Chinese broccoli, is a *B. oleracea* bred specifically for the tender stems with leaves and flower buds. It bolts quickly on the young shoot rather than making a head. It is sold as a commercial crop, as bunches of stems with leaves and flower buds attached. I grow it as an eat-all greens crop. It usually does not produce flower buds by the time I harvest it, but I don't care. The tender stems and leaves taste just the same with or without the flower buds, and are delicious. The water they cook in is also especially delicious. In addition, gai lohn is good in salads.

Evergreen Seeds, a specialist in Asian vegetables, carries nine different varieties. I have tried their 'Crispy Blue', 'South Sea', 'China Legend', 'Hybrid Blue Wonder', and 'Hybrid Southern Blue' as well as the more widely available 'Green Lance Hybrid'. They all looked and grew pretty much the same in my garden and tasted the same.

I sow the plants so as to end up with them spaced at about 3 inches (7.6 cm) apart in all directions. The plants take about ten weeks to produce a prime eat-all crop.

Gai lohn can be cut in chunks and boiled or added to soups or made into messes o' greens as I do with other eat-all crops. But I also sometimes boil or stir-fry the harvested stems with leaves whole and use them as a side dish. Because of the thick stems, I don't dry it. It's great frozen.

Gai lohn isn't as productive of biomass as some of the other greens. In addition, having a bigger proportion of stem probably means it has less nutritional value than the eat-all crops that are mostly leaf. However, gai lohn can be grown from late spring through fall in mild-summer areas, and has its own unique flavor. I like to produce at least one or two crops of it every summer.

'Groninger' Kale (*Brassica napus*). This is a same-species relative of 'Russian Hunger Gap' and 'Red Russian' and has their typical purple-tinted green leaves and purple veins. These other 'Red Russian'–type kales do not lend themselves to eat-all-style growing, however. Their central stalks are too tough and stringy, and the young plants have only sparse, deeply indented leaves and don't begin producing serious biomass of leaves until they are big established plants. 'Groninger' kale has apparently

been being grown and harvested essentially eat-all style for centuries. A friend tells me it overwinters in continental Canada.

'Groninger' is available from William Dam Seeds, a Canadian company that does not sell to the United States. I hope to make the variety available in the U.S. in spring 2016. The William Dam catalog calls the variety a collard or leaf cabbage, and says it is an heirloom "Old Blue Leaf Cabbage" from Holland that has been used for greens for centuries. However, it is clearly a *B. napus*, so calling it a cabbage or collard is misleading. Cabbages and collards are *B. oleracea*, not *B. napus*. *B. napus* varieties are usually called kales. So I'm calling it a kale.

The young plants of 'Groninger' grow faster than the other napus kales, have broader, less indented leaves and tender stalks, and produce a tasty eat-all crop. It overwinters well in the Willamette Valley. I don't have enough experience to know the acceptable planting windows. If 'Groninger' behaves like other kales, we might be able to plant it from midspring through fall.

Shungiku (*Glebionis coronaria*). Also known as garland chrysanthemum or edible chrysanthemum, this is a variety that is widely grown in Japan for edible greens. Unlike the rest of the eat-all greens, the flavor of the leaves boiled is quite strong, too strong to use as a stand-alone mess o' greens. It's better to use modest amounts of leaves in soups or stews. However, best of all is to stir-fry it. The flavor stir-fried is quite different from the flavor raw or boiled, and is spectacular. It's the signature flavor in the Japanese dish sukiyaki. I grow shungiku to use in stir-fries, and give it a miss as boiled greens. Shungiku is my very favorite stir-fried vegetable. It takes frying or braising about six to eight minutes before the characteristic suki-yaki flavor develops. The greens wilt and appear cooked several minutes before they develop the special flavor. There are narrow-leaved and broader-leaved varieties. The narrow-leaved types are usually stronger in flavor, but both make wonderful stir-fries.

Shungiku can be planted in midspring and then again in fall after the start of cool weather. Summer plantings bolt on young plants instead of producing good eat-all crops.

Shungiku takes about ten weeks to produce a crop and does not yield nearly the biomass of many of the other eat-all greens varieties. But just try it stir-fried, and you'll want to grow at least a little of it every year.

Amaranth (*Amaranthus* spp.). There are at least three species of domesticated amaranths and dozens of varieties of each. Some varieties are grown primarily for greens and some primarily for grain or as ornamentals. Leaf color can range from green to deep red, as can flower heads, depending on the variety. I've trialed more than a dozen varieties, in most cases in multiple seasons. Most of the types usually grown for leaves don't grow rapidly enough for good eat-all crops. 'Red Stripe Leaf', 'Redleaf', 'Emerald', 'Hopi Red Dye', and 'Polish' all do not grow anywhere close to fast enough to be good eat-all crops. 'All Red' is a bit too slow for a good eat-all crop too, but grows faster than the other red amaranths and is so spectacularly red that it jazzes up the entire garden. If you like amaranth in salads, it's worth growing 'All Red' for the color. 'Elephant's Head' and 'Love Lies Bleeding' are two varieties usually

grown as ornamentals that also do not grow fast enough to be eat-alls. 'Hopi Red Dye' has the most intensely red plants and leaves, but is so unvigorous it has a hard time even surviving in my garden.

The two varieties I've found so far that make the best eat-alls are 'Green Calaloo' (Baker Creek Seeds) and 'Burgundy', which is widely available. 'Burgundy' is actually a grain type. It has bronze-green leaves with reddish veins and burgundy flower heads. I'm continuing to trial different amaranth varieties. Ideally, I like green leaves, as colored varieties stain the cooking water.

Young amaranth greens are often used in salads. So in my variety trials I tasted all the amaranth varieties raw as well as boiled. Given the fact that there is such a range of colors and some varieties are traditionally grown explicitly for greens, I expected to get a large range in flavors or textures or quality. I didn't. Basically the raw leaves of all the varieties tasted the same to me. Blindfolded, I would have been unable to tell any of the varieties apart. Some of the leaf types might have had a little finer-grained texture than the grain types, but there was more variability from slightly younger or older leaves on the same plant than between varieties. All the amaranth varieties had leaves that were a little fuzzy and pretty dry. I don't particularly like any of them raw, even mixed with other things in salads. And when cooked, they also all taste the same. However, the red-colored leaves stain the cooking water purplish brownish black. So I prefer cooking greens that are straight green.

Amaranth greens are one of my favorite eat-all crops. They also make one of my favorite flavors of cooking water as well as tasty dried greens and herbal tea. And they are one of my favorites for freezing.

Amaranth is a warm-season crop. It can be planted from late spring after all danger of frost is over until late summer. I broadcast seed so as to end up with plants about 4 inches (10.2 cm) apart in all directions, and cut and use the top 8 inches (20.4 cm) when the plants are about 12 inches (30 cm) high. I sometimes clear-cut, and sometimes selectively harvest-thin to leave plants spaced in rows about 12 inches (30 cm) apart to produce seed. I sometimes also harvest-thin the plants, then top the remaining plants and let them turn into bushes that produce several meals' worth of greens from the young growing tips of the branches.

The 'Green Calaloo' does not produce enough seed to be worth growing for grain, but since it is available only from Baker Creek Seeds and only in packets with tiny amounts of seed, you need to do your own seed saving to use it as an eat-all. 'Burgundy' is generally available in packets and wholesale amounts. In addition, it produces nice grain heads and so can be considered a dual-purpose crop.

'Green Calaloo' and 'Burgundy' both grow quickly enough to make good summer catch crops. One of the photos in the photo section, for example, shows a few small patches of 'Green Calaloo' in a small corn-breeding patch where I had had erratic germination of the corn. Since I didn't want the amaranth to grow big enough to create shade and slow down the corn, I ate most of these amaranth plants at the eat-all stage (shown) and left just a few plants to produce seed.

CHAPTER ELEVEN

Peas and Beans

*Nitrogen Fixing and Legumes. Dry Seeds Versus Edible Pods Versus
Green Seeds. Pea Vine Types and Support. Shelling Peas. Edible-Podded Peas.
Growing Peas. Presoaking Legume Seed Without Suffocating It. Keep Peas
and Beans Picked. Harvesting and Using Edible-Podded Peas. Kinds of Bean
Varieties—Green, Dry, Shelly. Pole Versus Bush Green Beans. Seed Color
and Green Bean Flavor. Supporting Pole Beans. Growing Beans. Growing
Pole Beans on Corn. Harvesting and Using Green Beans.*

Nitrogen Fixing and Legumes

Plants need nitrogen. Air is full of nitrogen, but atmospheric nitrogen is in the N_2 form, which is quite inert and unusable by plants without some extra help. Peas and beans are legumes, and legumes have special relationships with various nitrogen-fixing bacteria. The bacteria are capable of converting the atmospheric nitrogen to ammonium compounds that plants can use. The bacteria and plants cooperate to produce nodules on the plants' roots where the bacteria live and supply usable nitrogen to the plant in exchange for sugar and other nutrients. If you dig up the roots of a pea or bean plant and wash the soil off the roots in a bucket of water you will find lumps on the roots that vary in size from tiny up to pea-sized. (If you pull the plants nearly all the nodules are stripped off the roots and are left in the soil, as are the smaller roots and root hairs where most nodulation happens.) The relationship between the nitrogen-fixing bacteria and legume plants is complex. For example, some bacteria try to cheat and nodulate legume plants and get free sugars without providing much nitrogen. But legume plants did not evolve yesterday. In the history of the universe no legume has ever been fooled into buying the proverbial Brooklyn Bridge. And they aren't fooled by would-be cheating nitrogen-fixing bacteria either. Legumes have mechanisms that allow them to identify and cut off the supply of sugar to nodules that are not delivering their quota of nitrogen.

Nitrogen is often the main nutrient we gardeners have to replace regularly with external inputs such as compost, seed meal, or manure. So vegetables with little or no need for externally supplied nitrogen are especially valuable.

Exactly which species of bacteria are involved in nodulating legume plants depends on the species of legume, and some bacterial species are better nitrogen-fixers than others. The nitrogen-fixing bacteria that associate with peas, cowpeas, soybeans, fava beans, and many vetches and clovers are generally much better nitrogen-fixers than those that associate with common beans (*Phaseolus vulgaris*). So it is cowpeas, soybeans, and various warm-season clovers that make the best summer nitrogen-fixing green manure crops, not common beans. And peas, overwintering clovers and vetches, and fava beans make excellent nitrogen-fixing overwintering cover crops.

Gardeners often think that they can interplant legumes and non-legumes so that the legumes will provide the non-legumes with nitrogen. It doesn't work that way. Plants do have some root connections with surrounding plants, and communication (or probably, more accurately, eavesdropping) does happen via these connections. However, there have been numerous studies done with radioactive plants that show that no significant amounts of nitrogen acquired by a nitrogen-fixing plant are transferred to adjacent plants. It is only when you till the legume under that the resulting dead legumes decay and release nitrogen to the soil, which then becomes available to surrounding plants or succeeding crops. So growing legumes saves you nitrogen-providing fertilizer because they don't usually need it. And they can provide nitrogen to the soil when they are tilled under. But you can't interplant nitrogen-fixing plants with other plants to provide nitrogen to other plants. (Exception: If you have a lawn or pasture with legumes that is mowed or grazed, legume greens are killed and recycled into nitrogen that is presumably made available to all the plants without killing or tilling under the whole legume plants.)

High nitrogen levels in the soil inhibit the growth and activity of nitrogen-fixing bacteria. So if you overfertilize legumes, they use the fertilizer you've made available in the soil and you get little benefit from the fact that the plants are legumes and could otherwise get their nitrogen from the bacteria, which fix nitrogen literally out of thin air.

Many gardeners these days think they need to inoculate their pea, bean, or other legume seed with store-bought spores of the right nitrogen-fixing bacteria in order to get proper nodulation of their peas or beans. Most seed companies sell little packets that contain a general pea and bean legume inoculant. These general garden inoculants contain a mix of bacteria appropriate for garden peas, common beans, runner beans, and lima beans. They may or may not include bacteria that associate with cowpeas and fava beans. You usually need to buy a different inoculant for soybeans. And another one for garbanzos (chickpeas). Exactly which species of legumes are covered depends on the brands and specific packet type.

For years I bought inoculants and inoculated most of every batch of peas or beans I planted. I always left at least a few feet of row uninoculated, however, as a control. There was never any difference between the inoculated and uninoculated sections with respect to growth rate of the plants or yield. And all the plants had roots that were abundantly nodulated, whether I inoculated or not. I've done this testing with peas, beans, cowpeas, fava beans, soybeans, and garbanzos, and I've done it over multiple years on at least five different soil types in five different gardens, some of which had been grass pasture with essentially no legumes for decades before the tests. Thinking that the little packets might have been stale and ineffective, I've even ordered garbanzo inoculant fresh from the factory to my door and used it within two weeks of its actual date of manufacture. It still made no difference. I've asked other organic gardeners and farmers around here, and many others have also done such trials. Nobody has seen any effect of inoculation.

Perhaps our maritime Northwest climatic conditions are distinctive enough so that the bacterial lines in the commercial inoculants don't thrive here and our legumes are all nodulated with strains of wild species. Perhaps if you have healthy organic soils you already have all the relevant bacteria you need. Perhaps you only need inoculants if you have poor, ruined soils. Maybe the whole inoculant industry is a scam. Or maybe the inoculants work in some sections of the country and not others. I don't quite know what is going on. However, at this point it's been years since I have bought any inoculants.

Inoculants are an external input you can't provide yourself. Part of resilient gardening is gardening in such a way that you could do without external inputs for a while if you had to. So it's very useful to identify unnecessary external inputs and eliminate them from your ordinary practice. In addition, inoculating legume seed is extra work, especially with large plantings. So if you have been inoculating, try doing uninoculated controls and figure out for yourself whether inoculating is giving you any benefits under your conditions.

Dry Seeds Versus Edible Pods Versus Green Seeds

Many varieties of peas and beans have been bred specifically for the dry seed. These "pulses" or grain legumes are very valuable staple crops because they have higher protein content than grains as well as good amounts of carbohydrate and are long storing. Dry pulses ought to be a part of your food repertoire even if you don't grow them. They are among the most valuable foods to stockpile. I think that every family should

have at least a few months of food stockpiled as part of their food resilience. Soup peas, cowpeas, and lentils are especially valuable for stockpiling because they can be cooked without presoaking. (Black-eyed peas are a cowpea, have a delicious beefy flavor, and are widely available commercially.) So if you are buying rather than growing your grain legumes, you might want to emphasize these. If you are growing your own pulses, what you can grow most easily and the yields matter more than the need for presoaking.

If you decide to store pulses and don't usually eat dry legumes, lots of luck if you suddenly try to cook them and eat them during hard times. Hard times are not made easier by needing to learn new food preparation and cooking patterns, or by the digestive system problems you get if you try to add serious amounts of dry pulses to your diet suddenly. Furthermore, while the dry pulses can last a year or two in storage, they don't last indefinitely. If they are part of your family's food resilience, they need to be part of your family's ordinary diet so you turn the supply over regularly, know how to prepare them, and are used to eating them. There is an entire chapter, a mini book, basically, on growing, threshing, cleaning, and cooking with dry beans in *The Resilient Gardener*. In this book I'll focus on growing peas and beans for their edible pods or green-stage shelling peas or beans.

Pea or bean varieties are usually appropriate for dry seed production or green pod production, not both. The very characteristics that make a legume variety good for dry seed production are the opposite of those that make it good for green pods. If we want both dry pea or bean seed and green pods, we grow different varieties for each purpose.

Pea Vine Types and Support

Pea varieties are usually described as short-vine or bush types, medium-vine, and tall-vine types, but the words *vine* and *bush* are both misnomers when applied to peas. A plant of a "vine" pea variety makes a single stem that grows straight and does no wrapping or climbing around other plants or adjacent supports. The plants support themselves by tendrils produced at each node that curl around any suitable supports they come in contact with, such as other pea plants or a chicken-wire fence support. Pea plants of the varieties referred to as "bush" aren't bushes either. A "bush" pea plant also produces just a single stem with tendrils at the nodes. The only difference is that the plant is shorter.

We usually crowd pea plants when we plant them. For example, I like to plant the seeds in wide rows about 4 to 8 inches across (12.2 to 24.5 cm), with the seeds spaced about 1 to 1.5 inches (2.5 to 3.8 cm) apart in all directions. Pea plants need to be crowded. In a wide row of pea plants the tendrils from each plant bind around the stems and tendrils of neighbors, and the resulting mat is easier to support than a single row of plants. It is also much higher yielding. With the shorter "bush" varieties, the wide-row mat is solid enough to provide all the support needed.

Medium- or tall-vine varieties need more support than is provided by a wide row of the plants clinging to one another. Most medium-vine types may grow up to about 4 feet (1.2 m) tall. Tall-vine varieties may grow up to 6 feet (1.8 m) tall. You can provide the necessary extra support by putting fencing along or adjacent to the wide row. If the fencing is adjacent to the row, the peas get support from the fencing mostly by virtue of the tendrils on the edge plants that happen to be

positioned just right to touch a wire and respond by wrapping around it. So fencing with small holes such as chicken wire better supports peas than does fencing with large holes and wider-spaced wires. When the fencing or support structure runs right down the middle of the pea row, the type and configuration of the fencing doesn't matter much, because the support structure is buried right in the middle of the pea mat and holds up the entire mat physically as well as via its connections to individual plant tendrils. Tall-vine types usually require serious tall support fencing. With medium-vine varieties there are various tricks we can use to make the job of supporting them easier than actually installing fencing.

These days I prefer making smaller succession plantings of medium-vine peas, and I make wide rows in a ring shape that is held up adequately by inserting a commercial tomato support ring in the center when the peas are up. (See the photo section.) Generally, medium-vine varieties produce much bigger pods and more pods than bush types. The difference is so great that I usually do not bother growing the short-vine types. Their advantage of needing no support is canceled by the ease of supporting my ring-patches of medium-vine types. The short-vine types may have shorter maturity times, however.

You can grow two adjacent rows of medium-vine peas and flop the two rows against each other just right so that the double row is self-supporting. Plant two 8-inch (24 cm) -wide rows with a gap of a foot (30 cm) between them. When the individual wide rows get to about 1.5 feet (46 cm) high they will be ready to flop over. You have to watch the rows and catch them at the right stage—when the rows are as tall as they can get without flopping over and are getting ready to flop in one direction or the other. Then you physically push and flop the two wide rows toward each other so they hold each other up.

A traditional pattern for supporting medium-vine pea varieties is to use "pea brush." You place brushy dry wood you have saved from pruning in the pea rows for support.

I usually add fencing, tomato rings, pea brush, or any support structures after the peas are up. This gives me a chance to do one good weeding without interference from the support structures.

I sometimes interplant later plantings of medium-vine peas with corn. Summer-planted peas seem to appreciate the shade in an established corn patch. I put a solid block of peas about 8 inches (24 cm) across in gaps within the rows where the corn plants didn't come up. The pea plants cling together to make a column that, with a little encouragement in the right direction, flops onto and is supported by one of the corn plants.

I used to grow some tall-vine varieties. I don't anymore. The problem is that tall varieties do not have bigger peas, better flavor, or greater yield than do the best medium-vine varieties. The tall-vine varieties don't have bigger leaves or more leaves than the best medium-vine varieties either. They simply have longer stems between the nodes so they grow taller and require taller supports. In addition, none of the taller varieties has the repertoire of disease-resistance genes that allow those of us in the maritime Northwest to grow peas all the way through the summer and fall.

Shelling Peas

Peas used to mean dry peas or soup peas. When people were talking about varieties intended for use as green peas, they called them "garden peas" or "shelling peas." Shelling peas are harvested when the pods are fat and the peas have swollen to full size but before they start to dry down. The dry seed is usually wrinkled rather than round like soup peas. The wrinkled seed marks the presence of a genetic mutation associated with sweeter flavor in the green peas and delayed conversion of sugar to starches in the seeds. The pods of shelling varieties are usually tough and have substantial strings that develop early. These pods and strings make it easier to shell out the green seeds.

These days I grow only edible-podded varieties. You get so much more food for the space and picking labor with edible-podded peas than with shelling varieties, and don't have to deal with the laborious shelling. And I like edible-podded varieties best anyway.

Edible-Podded Peas

Some edible-podded peas, called "snow peas," must be used when the pods are quite small, long before the peas begin to form. Snow peas are not at all sweet. "Sugar peas" usually may be harvested up until the pea seed begins to swell, and at that stage are sweeter than snow peas. Snow pea and sugar pea varieties have flat pods. Snap peas, which are round in cross section, have wrinkled seed and are harvested when the seed has fully formed but before it begins to dry down. With a snap pea the harvested vegetable includes both the succulent edible pod and the fully formed green peas inside. At this full-ripe stage, both the pod and the green seed are quite sweet.

My favorite pea variety is 'Oregon Giant Sugar', which was bred by Jim Baggett at Oregon State

University. It is usually classified as a sugar pod type because of its flat pods. However, like snap peas, the pod walls of OGS are thick, the pods stay succulent right up until the green peas inside are full-sized, and, if you harvest at this time, both the pods and peas are very sweet. We actually need another category for edible-podded peas, which I will designate "flat-snaps," for edible-podded peas that can be harvested when the pea seed is full-sized, that have thick, tender pods even at that stage, and that are very sweet at that stage, just like snap peas, but which have flat pods instead of round ones.

'Oregon Giant Sugar' has large wrinkled seeds and medium-sized vines that grow up to about 4 feet (1.2 m). The pods of OGS are the biggest and heaviest I've seen on any pea, including heirloom vine types such as 'Mammoth Melting Sugar'. Pods of OGS are up to 5.5 inches (14 cm) long and 1 inch (2.5 cm) across. And the flavor is as sweet and delicious as that of the very best snap pea varieties. By comparison, snap peas of even the best varieties are tiny. OGS produces two of its huge, heavy pods at nearly every node.

The leaves of OGS are large, allowing the plants to thrive in cloudy or overcast Oregon spring weather when direct sun can be rare. In addition, OGS is resistant to pea enation mosaic virus, pea wilt, and powdery mildew. This trio of disease resistances makes it possible to grow peas year-round in the Willamette Valley and to harvest peas from June into October. And the big leaves, fast growth, and great vigor make OGS an excellent choice for growing as an eat-all pea shoot variety, as I described in chapter 10.

I used to grow many different pea varieties, shelling, sugar pod, and snap pea types. I grow pretty much only 'Oregon Giant Sugar' at this point. If I harvest the pods at the snow pea size they are just like snow peas, tender and just the right size for stir-frying whole, and not especially sweet. If I harvest the pods at the sugar pod stage with the pea seed just starting to fill out, they are slightly sweet. And at full size with the seed filled out I get the most food for the space and growing and picking labor, and get that lovely snap pea sweetness and flavor. Even apart from the size of OGS pods, I find I prefer the flat-snap configuration to the traditional round snap peas. The flat pods work much better in stir-fries, and can also serve as dippers with which to scoop up tuna fish salad or cheese sauce.

'Oregon Giant Sugar', like most pea varieties including edible-podded types, has strings. There are some stringless snap pea varieties available. However, none has the trio of disease resistances needed for near-year-round production in the maritime Northwest. In addition, in peas, the stringless characteristic seems to be associated with poor germination, at least in current varieties. (I thank Jim Myers, Oregon State University vegetable breeder, for this information.) The strings on 'Oregon Giant Sugar' are relatively light and develop fairly late, however. If you pick the pods at the snow or sugar pod stage the pods are essentially stringless. I pick most of mine at the full-sized flat-snap stage. Sometimes I destring the pods as I pick them by snapping the pods off just below where the pea joins the stem, stripping the string off in the process. (It's often just the string on one side that is tough enough to bother removing.)

Nearly all edible-podded pea varieties have green pods. The heirloom 'Golden Sweet', often listed as a yellow snow or sugar pod type, isn't. It is a soup pea to which many people apply wishful thinking. The pods are beautiful, as are the golden plants with their purple flowers. You can eat the pods if you pick them when they are only an inch or two long (2.5 to 5 cm.), as you can with other soup peas. If you pick them at ordinary sugar pod size the pods are tough and stringy, just like other soup pea varieties.

If you want a true yellow snap pea variety with tender succulent pods and excellent flavor try 'Opal Creek', bred by Alan Kapuler (Peace Seeds).

There are many purple-podded soup pea varieties, and these occasionally get listed as shelling peas or even edible-podded peas. The only true edible-podded pea variety I know of with purple pods is 'Sugar Magnolia', a purple snap pea also bred by Alan Kapuler (Peace Seeds).

Growing Peas

Peas, like most legumes, don't require as much nitrogen as most garden plants but do care about sufficient calcium, phosphorus, and sulfur. Usually, soil that is adequate for most garden plants will be more than adequate for good pea crops.

For early-spring plantings, presoaking the pea seed can be useful. This breaks the dormancy of the seed under warmer indoor temperatures. The presoaked seed then needs less warmth to continue growing than it would if you were to just plant dry seed outdoors.

If you use inoculants on your pea seed, first sow some uninoculated seed as your control before you start treating seed with nitrogen-fixing bacterial inoculant, as I mentioned in the section Nitrogen Fixing and Legumes. You may find that under your conditions inoculants make no difference and that you can dispense with them in future.

Just sprinkling inoculants into the row or onto dry seed is not effective. Put the seed and inoculant together in a container just before planting and add enough water to turn the inoculant into a thin mud that coats and sticks to all the seeds. Wet seeds clump together and are hard to handle and sow. However, if you add just the right amount of dry soil to the seed and inoculant mud the soil will dry off the seeds enough so that they will become separated and easier to plant. Sow seed immediately after inoculating.

Peas love cool weather, and the young plants are capable of handling mild to hard freezes, depending on the variety. Peas hate hot weather, however. So in most regions of the United States and Canada, peas are planted as early as the ground can be made ready in spring and are harvested in summer. In the South and in New England it may also be possible to plant again for a fall crop.

The maritime Northwest is Pea Heaven. Our rather cool summers allow for a much more extended pea season than can be enjoyed elsewhere, but only with certain varieties, those resistant to pea enation mosaic virus, pea wilt (*Fusarium*), and powdery mildew. Varieties with this triplet of resistances that can be used for extended pea harvests in the maritime Northwest are 'Oregon Giant Sugar', 'Oregon Sugar Pod II', and 'Oregon Trail' (a shelling pea), all bred by Jim Baggett at Oregon State University, and all released as public domain varieties. With these varieties we can plant our first planting of peas in February or March and our last planting in the first week of August and

Presoaking Legume Seed Without Suffocating It

It is often handy to presoak pea or bean seed. I usually presoak my earliest spring planting of peas because it may rot if the weather doesn't stay warm for long enough to germinate it. Early-spring weather is treacherous. Presoaked pea seed can be planted earlier than unsoaked seed because it takes more and longer warmth to break dormancy and initiate germination than it does for germinated seed to grow. I get better stands from the presoaked seed. I also usually presoak my fall-planted peas because I am often planting where I have no irrigation set up at that point. I usually don't presoak my summer-planted peas, because neither the weather nor the water is critical. I usually don't presoak beans except for those I interplant with corn. I presoak those because the corn is already up and has big enough root systems so that I don't need to water it very often. I presoak the bean seed so I don't have to water the entire corn patch more frequently just to germinate the later-planted bean seed.

If you just toss the legume seed in a container of water to presoak it, most of the seed will suffocate and die from lack of oxygen. Seed uses lots of oxygen as it imbibes water and starts to germinate. In order to properly presoak legume seeds we need to provide both water and oxygen. A traditional way of presoaking seed was to put it in a cloth bag, tie off the bag, then hang it in a creek for a couple of days where the water was constantly replaced with fresh water that was fully oxygenated.

With amounts of seed up to quart or so, I put the seed in a very large bowl or a bucket that holds

---

can harvest peas throughout summer and fall until the first freeze of winter. Varieties without enation resistance, wilt resistance, and powdery mildew resistance should be planted in spring only. To my knowledge, no heirloom variety carries resistance to the critical three diseases. A single planting of peas usually bears pods for only about three weeks. So I usually plant a new pea patch every month or so from spring through the first week in August.

In the maritime Northwest we can also fall-plant and overwinter many pea varieties, including 'Oregon Giant Sugar', 'Oregon Sugar Pod II', and 'Sugar Snap'. To overwinter peas plant them at the same time as is recommended for cover crop peas in your area. (The first half of October is the right time in the middle Willamette Valley.) Overwintered peas do not give earlier harvests in spring. Instead, they give you much larger flushes of early peas, as the plants are large and well established when the weather warms up enough to permit flowering, and they flower and set pods on several nodes on each plant all at once. The major reason I overwinter peas is because it is usually easier for me to start peas in October than in the unpredictable weather and mud of February or March.

In my experience most medium- and tall-vine varieties with big leaves overwinter in the Willamette Valley, but most short-vine types and varieties with small leaves do not. Short-vine types usually get eaten by slugs because the several gallons of water. Wide containers are better than deep ones, as there is more surface area for the water to take up oxygen from the air to replace what is depleted by the seeds. But active stirring and complete replacement of the water occasionally is also needed. I start with cold water for the first hour or so. (A slow start is less likely to make the seeds split.) Then I replace the water with warm water. Then I stir the seed at frequent intervals during the next day or two, as well as replacing the water completely with fresh water at least three times per day. (Call it at least six stirrings per day, three completely replacing the water as well.)

When the seed is ready to plant it will be completely swollen and the radical, the little root of each seed, will be visible but will not have broken free of the seed skin. At that point the seed has broken dormancy, has imbibed all the water it needs for germination, and in fact has actually initiated germination. In another day the radical will break free of the seed coat, making it vulnerable to damage in planting. At this optimal stage, with the radicals still protected by their seed coats, the seed can be handled just like dry seed. (Once the radicals have emerged you have more of a transplanting situation and a huge increase in labor.) I mix the presoaked seed with enough dry soil in the garden to dry the seed and make it free flowing and easy to handle.

Good viable seed of pea and most common bean varieties usually needs about twenty-four hours to soak to the perfect stage at ordinary indoor temperatures. Fava beans, garbanzos, and runner beans may take two days, depending on variety. If your batch of presoaked seed has many unswollen seeds you either have seed with poor viability or did not give the seeds enough oxygen during the soaking.

apical tips stay down within prime slug foraging height for longer than do those of taller varieties. Small-leaved types probably can't photosynthesize enough to make it through an almost totally overcast Oregon winter.

Peas like it crowded, and I usually like to plant peas in wide rows as described in the section Pea Vine Types and Support. I make a shallow trench about 1 to 1.5 inches (2.5 to 3.8 cm) deep and about 4 to 8 inches (12 to 24 cm) wide using my peasant hoe. I sprinkle the seeds in the trench, then gently hoe the soil back over the peas to cover them. You don't need to be accurate or uniform with spacing. I am after an *average* spacing of 1 to 1.5 inches (2.5 to 3.8 cm) apart in all directions, but I don't get in there and arrange or rearrange the seeds by hand. I just sprinkle them and leave them where they land. Some peas will land right next to each other. I don't worry about it, and neither do the peas.

As for choosing appropriate varieties, I would suggest that, wherever you live, try 'Oregon Giant Sugar' and 'Oregon Sugar Pod II' in addition to whatever varieties are the most popular in your region. The former is my favorite. The latter used to be my favorite before OGS was released, and it is the most widely grown sugar pod type in the world, so is pretty widely adapted. Different regions have different diseases that can matter, and in areas with hot summers, heat tolerance is an issue. So consult regional seed catalogs for varieties that are adapted to your area.

Keep Peas and Beans Picked

We want the pods or green seeds, but pea and bean plants are striving to make dry seed. When we leave some pods on a plant, the plant stops making more pods and focuses its energy on maturing and drying down the pods it has already made. So if you leave a few pods on the plant, you are not wasting just those pods. You are taking the plant out of production permanently. Clean-pick your peas and beans. That is, remove any pods that escape you and are past prime; don't leave them on the plants. If you find you have more pea or bean patch than you can keep properly harvested, focus on the best section and keep that completely harvested so it continues producing, and write off the rest.

Harvesting and Using Edible-Podded Peas

Snap pea varieties (including the flat-snap 'Oregon Giant Sugar') must be allowed to stay on the vine until the green seeds are full size for optimal sweetness. The earlier you pick them before that stage, the less sweet they'll be. If picked before the seeds start to swell, the pods will not be sweet at all, though they are still tasty. Once the green peas inside the pod have reached full size, the pods must be picked before the peas start converting sugar to starch or drying down. An added reason why I appreciate 'Oregon Giant Sugar' is that after the pods develop full-sized seeds they seem to just sit there for a few days without beginning to make starch. Thus there is a longer harvesting window than for most sugar pod or snap pea varieties.

Here in maritime Oregon, I've noticed a big difference in the sweetness of peas depending on when they are picked, especially for the early-spring plantings. Pods picked in early morning or throughout the day on cool, overcast days are not as sweet as those picked later in the day on a day when the plants have had some sun. The difference in sweetness is large and obvious, not subtle. I don't know whether the difference in sweetness with harvesting time and conditions is specific to my climate, however. I suggest that you learn to identify the sweetest stage for picking your chosen variety or varieties by picking and eating pods in the garden. Then, once you have sorted out that factor, try picking at different times of day and see if it matters.

I pick peas into plastic buckets or containers, not metal ones that become hot in the sun. I pick just before leaving the garden, cover the container after picking, and bring the pods straight home and put them into the refrigerator promptly in sealed plastic bags. (Except for the big bowl of pods I usually set aside to eat immediately.) If handled in that way, the pods will keep in the refrigerator for at least three or four days with no signs of deterioration or loss in flavor.

My very favorite way to eat 'Oregon Giant Sugar' peas is raw. I eat large amounts of them just straight. Sometimes I use them as scoops to dip up tuna fish salad or a cooled cheese dip. I also really love them in tuna fish salad and all other salads.

I chop up the pods and add them to soups and stews during the last two minutes or so of cooking. I usually use as much pea pod as all the other ingredients put together, and the soup or stew has a wonderful distinctly pea flavor. Occasionally in reheating a soup I overcook the peas. In this case, they lose some of their color and end up tasting like canned peas. But that is okay too. I also occasionally like that flavor.

Kinds of Bean Varieties—Green, Dry, Shelly

Some bean varieties are intended for use as dry beans. For these, it is the flavor and production of the dry beans, not the flavor, characteristics, or yield of the green pods or green seed that matter. Plants of good dry bean varieties usually have a growth pattern that optimizes for dry bean production but makes it difficult or impossible to get much of a harvest of green beans or shellies. ("Shellies" is what we call full-sized but green bean seed. It is the bean version of what we call "shelling peas" in pea varieties.) After growing to full size, each pod of a dry bean variety quickly rushes into swelling out and maturing its seeds, then quickly rushes into drying down those seeds. The bean pods pass through the point at which they are full-sized but still tender almost instantaneously, meaning it is very difficult to find many pods to eat as green beans, even in a large planting.

Furthermore, with bush dry bean varieties the pods are often small. Pod size doesn't matter for dry bush bean varieties because we usually don't pick the pods. We thresh the beans from the dry plants with the pods on them. Whether we do it with a combine in the field or by cutting the dry plants, piling them on a tarp, and stomping on or driving over them to thresh out the seeds, the pod size makes no difference. (For pole dry beans these factors do matter since we must usually pick the pods by hand.)

The bean pods of good dry bean varieties are also stringed, which makes them easier to thresh. The pods pop open easily along the strings. I grow lots of dry bush beans for eating as well as selling, including some I've bred myself—'Hannan Popbean' garbanzo, 'Fast Lady Northern Southern Pea', 'Beefy Resilient Grex', and a reselected line of the heirloom 'Gaucho'. There is a large chapter on growing and using dry beans (but not green beans) in *The Resilient Gardener*. So in this book I will focus on green and shelly beans.

In good green bean varieties it is the flavor and characteristics of the green pods that matter. A variety whose pods turn an unattractive gray or become soft and soggy when cooked is not as desirable as one that holds its texture and color through cooking. And since we usually pick the beans by hand, and bigger beans mean more food per picking gesture, for green beans the size of the pod really matters.

Good green bean varieties usually have a different growth pattern from those bred for dry beans. In most varieties, the pods toughen as the bean seeds inside develop. So in green bean varieties, the beans grow to full size, then the beans inside develop slowly. This means each pod is at the right stage for eating as a green bean over a period of several days, and each plant usually presents several pods that are ready to pick for green beans at once. With modern varieties, the toughening of the bean pod may also be delayed until later in the development of the seeds. This makes for a very extended harvest window for each pod.

Green beans used to also be called "string beans," and laborious destringing of each bean in the kitchen was a required part of preparing beans. Not any longer. In modern green bean varieties the pods are stringless. Beans of stringless varieties germinate as well, taste as good, and produce as well as those of stringed varieties. These days, people who buy beans in the market really do not expect to have to destring them, so if you are a market gardener, you will necessarily be growing modern stringless varieties.

I'm a big fan of heirloom dry beans. But when it comes to green beans, with a few exceptions I prefer the modern varieties. I am just not willing to spend all that time in the kitchen destringing beans. In addition, most of the heirloom green bean varieties produce pods that are only up to about 6 inches (15.2 cm) long at best, with 3 to 4 inches (7.6 to 10.2 cm) being the norm. A number of modern green bean varieties produce beans that are tender and tasty up to about 11 inches (28 cm) long. In some the pods are tender even after the beans inside are fairly large. This gives an extraordinarily long harvesting window. In addition, modern varieties often have useful resistance to various diseases not contained in the heirlooms that can add up to bigger and more reliable crops. These days I grow heirloom green beans only when they are of the most spectacularly flavorful varieties (such as 'Kentucky Wonder') or of special types bred for purposes that modern varieties don't have, such as the shade tolerance of traditional cornfield beans.

Some bean varieties are used primarily for shelly beans. Most of these also produce good crops of dry beans, but it is the flavor of the shelly bean that is special. Good shelly bean varieties grow their pods and fill them out rapidly. Then they wait around awhile before beginning to dry down their beans, making for an extended harvest window for the shelly bean stage. Shelly beans have stringed pods so that they are easy to break open to remove the beans. The pods are not edible at the stage they are picked for shellies. You can sometimes find them in farmers markets as somewhat leathery-looking fat pods filled with beans.

Most lima bean varieties, most southern pea (cowpea) varieties, and most large-seeded fava bean varieties are shelly bean types. Of the common bean (*Phaseolus vulgaris*) varieties, most are green bean or dry bean varieties, but there are a few that are shelly bean varieties. Some are 'French Horticultural', 'Wren's Egg', 'Bird's Egg', 'Vermont Cranberry', 'Specked Bays', 'Flageolet', and 'Cannellini'. These all have distinctive delicious flavors in the green stage but are relatively uninteresting as dry beans. (Exception: 'Flageolet' also tastes distinctive and delicious as a dry bean.)

Runner beans (*Phaseolus coccineus*) may be best as dry beans, green beans, or shelly beans depending on variety. 'Scarlet Runner', often grown as an ornamental, is one of the most delicious of dry beans. 'Blackcoat Runner' is another delicious dry bean type. 'Scarlet Emperor' has long pods, delayed pod maturing, and few seeds in the pod and is for green beans. And some varieties (usually white-seeded types with huge bean seed), such as 'Jack in the Beanstalk' aka 'Jack's Bean', are for shelly beans. Runner bean varieties meant for green beans are apparently more popular in Britain and Europe than in North America. I don't especially like the flavor of the pods compared with those of common beans. No one else around here does either. And we can grow them easily given our mild summers. In many sections of the country runner beans don't produce until late summer or fall because they are much more heat-sensitive than common beans. I consider runner beans best for dry beans or shellies, not for green beans.

I have only limited experience with fava beans as shellies. I am just not up for all the work involved in hand shelling them. Fava bean pods are difficult to open. The halves don't split open readily, despite the strings. The seeds are big, but you have to fight hard for each one. And most fava beans aren't that great as dry beans. Until I can develop a variety that is delicious as both a dry bean and a shelly, and overwinters well, I'm skipping growing favas. (I'm working on it.) I have started experimenting with runner bean shellies; I'm enchanted with the huge seeds, biggest of all the edible bean seeds by far. And they are mostly pole varieties I can grow on corn. These days, though, when it comes to green beans or shelly beans, everything has to meet the WIRHaMoOGS standard. In other words, would I rather grow this, or "Would I rather have more 'Oregon Giant Sugar'" edible-podded pea? Given the fact that I base my books primarily on personal experience and my own in-garden research, this means you will need to go elsewhere for anything further about shelly beans. About the best I can do is tell you to grow a shelly variety only if you want shellies. There are better varieties if you want green beans or dry beans. And cook shelly beans well. Dry, uncooked bean seed is poisonous enough so that just three or four can make you seriously sick, and we don't know exactly when the anti-nutritional chemicals are synthesized during seed development. So cook shelly beans. And don't eat green beans raw if the seeds inside have started to form.

Pole Versus Bush Green Beans

For many years I grew only pole green beans. This was for several reasons. First, with my early gardens, space was at a premium. Pole beans are much more productive per unit space than bush beans, and the difference is large, two-fold or more per foot of row. Pole beans have an extended harvest. Bush beans usually yield for about three weeks, so you need several plantings for a full bean season. Third, some of the pole bean vari-

eties produce really large beans, beans that give far more food per picking gesture than the beans from bush green bean varieties. The biggest-podded pole bean varieties generally have pods up to about 11 inches (28 cm); bush bean varieties usually have pods up to only 6 inches (15.2 cm). (An exception is the bush bean variety 'Jumbo'.)

Most of the beans on pole varieties can be picked standing up rather than stooping over or kneeling.

Finally, pole beans were reputed to have the most flavor. At this point I'm not sure there is such a simple correlation, though. I think this impression may have been created by comparing certain very powerfully flavorful pole bean varieties to the blander commercial types of bush beans rather than comparing the most flavorful of each type. However, I know of no bean with a richer, meatier flavor than 'Kentucky Wonder', a brown-seeded heirloom pole type. A handful of KW beans in a pot of soup is enough to flavor the whole pot. Milder-flavored beans can be delicious too, however. And there are plenty of mild-flavored pole varieties. 'Blue Lake' pole is a milder-flavored classic that is many people's flavor favorite. There are good 'Blue Lake' bush varieties available, but most people think they really don't quite measure up to the pole variety.

At this point I think specific variety has more to do with flavor than whether the variety is a bush or pole type. There are plenty of commercial green bush beans that don't have much flavor, but nobody is holding a gun to your head and making you buy them. You can buy the home garden varieties that are noted especially for good flavor, such as 'Provider', for example. I attended a bean tasting last summer in which we sampled a dozen or so different varieties, both bush and pole types. They all tasted different from one another and from every other bean I had ever eaten. And I liked them all. I would have been happy to eat a big batch of any of them.

The earlier bush bean varieties are earlier than pole bean types. So some committed pole bean growers still grow an early planting of an early bush variety to give them beans until the pole beans start bearing. When I am growing a corn variety that is good for bean support, I usually plant pole beans for the main and late crop and bush beans for the early crop. If I'm not growing a corn that is good for bean support, I tend to plant more bush beans.

If you are in exploration mode on bean varieties, I would suggest trying 'Kentucky Wonder', which is very widely adapted, and 'Fortex', one of the modern pole bean varieties that is full-flavored and good to eat up to about 11 inches (28 cm). And also try 'Provider', an early bush bean with good flavor that is widely adapted, and 'Jumbo', the bush bean with the biggest pods (up to 11 inches/28 cm). Then of course also try whatever varieties are most commonly grown by gardeners in your region.

Seed Color and Green Bean Flavor

Many modern green bean breeders focus primarily on white-seeded varieties. When a green bean with brown or black seeds is cooked, the immature seeds inside the bean stain the cooking water. This is considered horrible for canned beans. It doesn't matter at all for home cooking if you throw away the cooking water. I don't. However, I'm perfectly happy to use the stained cooking water in my soups and stews if it contributes a rich beany flavor. I prefer unstained cook-

ing broth only if I can get it at no cost. There is real cost to white-seeded bean varieties, however. Both germinating vigor and flavor of the green pods seems to be linked to or influenced by bean seed color. Beans of white-seeded varieties of common beans usually have lower and/or less vigorous germination rates than those of colored varieties. In addition, the green beans of white-seeded varieties usually have mild flavors. All the richest, most powerfully flavored green bean varieties, without exception to my knowledge, have colored seeds.

Supporting Pole Beans

Pole bean plants send out a number of runners that are true vines that twine around other plants and support structures. It's interesting to watch them do it. Each vine will twist around in the air over a period of a few days, circumscribing a circle a foot across or so until it hits something of an appropriate diameter that is oriented vertically. Then the vine twines around that something. To grow pole beans you need to provide support. If you don't, they just flop and grow on the ground and produce little, most of which is dirty, moldy, and inedible. Trust me. Support your pole beans.

Most bush bean varieties are actually true bushes. They have relatively thick central stems and make a number of branches, and the plant stands up well enough to hold the pods up off the ground.

Some varieties that are called bush types are actually short-vine types. If these hold the beans up off the ground you can grow them without support just like true bush types. If instead they drag their beans on the ground, they will need support. 'Jerusalem Runner', a "bush" runner bean grown for its shelly beans, is a short-vine type that drags its pods on the ground and needs some support.

A traditional way to support pole beans was on poles. An 8-foot (2.4 m) pole about 6 inches (20 cm) in diameter with the bark left on is ideal. Some pole beans grow much higher than 8 feet (2.4 m); however, the beans higher on the plant are smaller. Too small to make it worthwhile putting up higher supports and then climbing on a ladder in order to pick the small beans above the easy picking level. So as far as I am concerned, supporting the first 8 feet of a pole bean variety is usually good enough. Bury one end of the pole in the ground, then plant about six to twelve pole bean seeds around the pole and thin to six plants. I sometimes use a metal T-pole because it is easy to pound into the ground, then tie a wooden pole to the T-pole for extra width and height.

Pole beans can also grow up corn or sunflower plants, trellises, or fencing. A common way to support pole beans is with strings. Put sturdy posts at the ends of the row, then attach a heavy wire or cable from the top of one pole to the top of the other and another cable between the posts near the ground. Then run heavy string around the two cables, looping it around the wires about 6 inches (20 cm) apart to provide nearly vertical strings for the vines to climb. Then plant a single row of pole bean seeds on each side of the support. To a bean plant, "support" means something vertical they can wrap around to climb upward. They don't care about and can't use the horizontal wires in a fence support.

The bean tepee is another classic support method. Tie three 2-inch (5 cm) poles together at one end and spread the other ends out, then insert them into the ground to make a tepee shape. Then plant the pole beans in a ring around each pole. Note that the poles must be at a fairly steep upward angle. If, for example, the poles are at a 45-degree angle with the ground, the bean vines won't be happy. They twine around things in order to grow upward, not sideways. The vine tips will desert the pole and start swinging free and searching for better support, and when they don't find any will droop down unsupported.

I often support pole beans with a section of cattle panel. It's only 4.5 feet (1.4 m) high, but I interweave sticks (yard debris) into the top couple of sections to extend the effective height.

For growing pole beans on corn see the separate section below.

Growing Beans

The common bean (*Phaseolus vulgaris*) is a warm-season plant that can't take frosts. And they don't grow well in cold weather. Runner beans (*P. coccineus*) are slightly more cold-tolerant and considerably less heat-tolerant. Lima beans (*P. lunatus*) generally need more warmth than common beans. All of them are best planted only after all danger of frost is over. Bush bean varieties often yield for only about three weeks, so multiple plantings are necessary to have beans all season. Pole bean varieties don't usually bear quite all season, but if we plant an earlier and a later variety we can cover all but the earliest part of the bean season. For the earliest bean harvests we need an early bush bean variety.

For information on soil needs, how to inoculate if you are using inoculants, or how to presoak the seed, see the section on Growing Peas.

I plant bush beans in single rows with about 3.5 feet (1.1 m) between rows. I make a furrow about

2 inches (5 cm) deep with my hand-furrowing plow, or with a furrower or plow attachment on a wheel hoe. A peasant hoe or other heavy hoe will also do the job. Then I walk along the furrow sprinkling the seeds in. When I cover the seed it is 1 to 2 inches (2.5 to 5 cm) deep. I usually sow heavily, then thin to 4 inches (10 cm) apart when the plants are at least about 4 inches (10 cm) high—beyond the stage at which cucumber beetles are likely to wipe out the seedlings. Occasional bean plantings are lost entirely because the seedlings are emerging just as the overwintering cucumber beetles emerge. Usually just replanting will work.

I plant pole beans in two single rows on either side of a support structure. (I plant the beans first, then put up the structure.) I let the plants get to 4 inches (10 cm) tall, then thin them to 6 inches (15 cm) apart in the rows. If I'm planting around poles or on a bean tepee, I always plant and thin so as to give the plants 6 inches each. This is a case where uniform spacing matters and crowding is detrimental. Pole bean plants are huge plants that produce high yields. They need the space.

Growing Pole Beans on Corn

To interplant corn and pole beans, first choose a corn variety that has enough height and a strong enough root structure. This usually means full-season field corns of traditional varieties that are 6 feet (1.8 m) tall or taller. Hybrid sweet corns usually have root systems that are too wimpy to grow beans on. Traditional open-pollinated late sweet corns often work, however. Early corns aren't big enough. Field corns are better than

sweet corns, but with care you can harvest most of the sweet corn ears despite the beans. My main food corns are ones I've bred myself—'Cascade Ruby-Gold Flint' and 'Cascade Creamcap Flint' for gourmet cornbread and polenta and to be the ultimate survival crop you would want for getting by in hard times. And 'Magic Manna Flour' corn for parching corn, cakes, cookies, pancakes, gravy, crackers, and cornbread. (I sell these through my company Fertile Valley Seeds.) I don't grow pole beans on these corns, though, because the plants aren't tall enough, and they are done in August. We quit watering the corn to help it finish and dry down just when pole beans would need water the most.

My favorite variety of corn to grow pole beans on is 'True Gold', an 8-foot (2.4 m) tall open-pollinated late sweet corn bred by Alan Kapuler by dehybridizing 'Golden Jubilee Hybrid' (the old standard, not the supersweet version). ('True Gold' is available from the Kapuler family seed companies Peace Seeds and Peace Seedlings, and from Fertile Valley Seeds.)

Match the pole bean and corn varieties. Don't try to grow a 20-foot (6.1 m) high pole bean on a 6-foot (1.8 m) high corn. You can, however, grow a 12-foot (3.7 m) high pole bean on an 8-foot (2.4 m) high corn variety. The reason is that the bean vines are a bit fickle in their commitment to individual corn plants. The vines grow up a corn plant for a while, then desert it and find another and grow up it for a while, and so on. So some of the vine length is used up in running sideways. This means 12-foot (3.7 m) tall beans usually end up only a little taller than 8-foot (2.4 m) tall corn. This pattern the beans have of using multiple cornstalks helps bind the row of corn together and makes the cornstalks less prone to lodging (falling over) than the same variety grown without beans, especially when the corn is growing in hills.

There are dozens of varieties of heirloom beans with *cornfield* in the name that traditionally have been grown on corn. Some of these are unusually shade-tolerant; others are no more shade-tolerant than most pole beans. (All beans will *grow* reasonably well in partial shade, but growing is not producing beans; most varieties produce much better in full sun.) Southern Exposure Seed Exchange has a good list of traditional cornfield bean varieties. Some of these varieties are best for dry beans, and these usually have to be picked a pod at a time; the pods dry down over a long period and are vulnerable to rain. So for pole dry bean varieties we usually pick all the dry pods before each rain or irrigation. The runner beans 'Scarlet Runner' and 'Blackcoat Runner' are excellent as dry beans, can grow happily on corn, and the pods are not as vulnerable to rain when dry as common bean types. And the seeds are huge, making the handpicking more worthwhile. Other cornfield varieties are best for shellies and others for green beans. I prefer to grow green bean varieties on my corn and bush beans elsewhere in a patch of their own for dry beans. 'Withner White Cornfield' bean, available from Peace Seeds and Adaptive Seeds, was long my favorite for growing in corn. It is productive in nearly full shade. However, these days I usually grow modern pole varieties such as 'Fortex' for the bigger and stringless beans and the heirloom 'Kentucky Wonder' for its special meaty flavor. Any pole bean will do fine in a corn patch if you either grow it on just the edge rows, or if you leave some extra space between corn rows with gaps here and there to let more light into the patch, as I described in the section Polycultures in the tomato chapter.

Most people who try to grow pole beans on corn fail for one of two reasons. The first is that they plant far too many beans per corn plant. The correct ratio is one pole bean plant for every three or four corn plants. If you plant too many beans the beans will either cause so much shade that the corn doesn't thrive, or they will create so much weight that they knock over the corn.

The second way in which people often go wrong in trying to interplant corn and pole beans is that they plant them at the same time. If you do that, the beans overgrow and shade the corn and may also knock it over. You have to give the corn a head start. Plant the corn first and let it get about 4 inches (10.2 cm) high. Then thin the corn to final spacing and plant the beans. I always presoak bean seed when I plant it with corn so that the water needs of the germinating beans will match the water needs of the already established corn.

I saw a recent article in a popular magazine on growing beans that described how to grow pole beans on corn. The author's instructions included adding belated stakes to the field to support the beans after they knock the corn over. That is how *not* to grow beans on corn. That author had clearly chosen a corn variety that was too wimpy to hold up beans, or a bean variety that was too tall for the corn variety, or had planted too many beans for the amount of corn. When you are doing it right, the beans do not knock the corn over, and you don't have to give them any support other than what they get from the corn.

There are three or four basic patterns for arranging the pole beans and corn. In the traditional Southwest Indian pattern, the corn varieties were usually tall and were planted in hills 2 or more yards (1.8 m) apart because of the scarcity of water. Each hill would have three corn plants and one bean plant. The bean plant wound around all the corn plants and bound them together and helped support them from lodging in sometimes fierce winds.

European pioneers usually grew their corn in rows, and fertilized so as to allow higher planting densities than in most Indian gardens. Where the corn is densely planted, beans can be grown on the sunniest edges of the patch. Alternatively, special shade-tolerant cornfield varieties can be grown throughout the patch.

At this point, I grow tall late sweet corn as much for the pole bean crop as for the corn. So I am willing to increase the space between the rows from 3.5 feet (1.1 m) to 4.5 feet (1.4 m) as well as to make occasional gaps in the corn rows to let more light in for the beans.

Note that if you try to skimp on the distance between the corn rows, the beans will sometimes twine between plants on adjacent rows and make it impossible to walk down the paths without cutting the errant bean vines.

With my standard 3.5-foot (1.1 m) spacing between my corn rows, I plant the beans in gaps within the corn row. Part of the reason I increase the spacing to 4.5 feet (1.4 m) is because that gives me enough room to plant the beans in a row adjacent to each corn row without blocking the path. It is easier to plant the beans adjacent to the corn row, which I can do standing up, making a ditch, and mass producing. Planting within the corn row after the corn is up is more a matter of getting down on hands and knees and individually hand planting each gap in the corn row.

I plant each pole bean row north of the adjacent corn row. This makes it easier for the bean vines to find the corn plants, since the beans are phototropic (grow toward sunlight) by default.

Harvesting and Using Green Beans

I harvest green beans exactly the same way I harvest peas, with the plastic bucket, beans covered on the way home, then refrigerated in sealed plastic bags promptly. However, with beans I have not discovered any flavor difference based on picking at different times of day, as I have with peas.

The right stage to pick the beans depends on the variety and your personal preferences. I prefer to pick the beans when they are as big as possible before the pods become tough. They can also be picked earlier, and the flavors can be different at different stages. So it is worthwhile experimenting with picking at different stages for each variety you try.

With 'Kentucky Wonder', for example, I let the beans get full-sized but pick them before the seeds inside develop at all. At this stage the beans are up to about 5 inches (12.7 cm) long and are stringless. If you wait until the beans have started to fill out even a little the pods become tough and the strings develop. With the modern pole bean variety 'Fortex', however, the pods stay tender and stringless right up to when the beans inside are pretty far along. I usually pick beans about every three days, so I pick beans that are full-sized or a bit smaller, that is, about three days short of full-sized to full-sized. So for me a mess of beans often involves mixed sizes and stages. When I cook beans of mixed sizes I usually snap the ends off the bigger ones and toss them in the pot first while I'm preparing the rest so that the more mature beans get cooked a little longer.

The cooking time affects green bean flavor a lot. 'Kentucky Wonder', for example, needs to be cooked eight minutes or more before it develops its rich, meaty flavor. It doesn't taste like much if cooked only for a couple of minutes. I prefer most varieties boiled for just a couple of minutes, though, in unsalted water. Most varieties don't taste like much if boiled as long as 'Kentucky Wonder'. However, 'Blue Lake' pole was the mainstay of the canned bean industry (before bush beans and machine picking took over) because it holds its color, texture, and flavor even when canned. When you are new to a variety I suggest tossing a few beans whole into a pot of boiling water, then tasting them at two-minute intervals until they are obviously overcooked.

The flavors of green beans seem to be more concealed than enhanced by salt. I prefer to eat my green beans unsalted.

You can steam instead of boiling your beans, but steaming always takes longer and gives a less uniform result, and is limited to smaller amounts. I often drop green beans into soups or stews. Some people bake them in casseroles. Pioneers ran strings through them and hung them near the woodstove for dried green beans for winter. Green beans can also be stir-fried. I usually just boil them or put them in soups or stews. Once boiled, you can also dress them in the various ways I described for eat-all greens. A classic southern dish was green beans dressed with or cooked with bacon, bacon fat, or salt pork and seasoned with pepper.

I prefer to cook green beans whole rather than cut or snapped into sections. They retain more flavor if not cut up before cooking. As a side dish I eat the green beans whole. In soups or stews I slice the beans up after cooking as they go into the individual bowls.

道

Fill yourself with utmost emptiness. Embrace internal tranquility.
The ten thousand things, see how they arise and flow around you—
each one coming into being, growing, adapting, changing, fulfilling,
then returning to the source—as you sit in stillness in the center, watching.

Be aware, observe, notice. Appreciate the small.
Find flavor in the delicate.

Give birth without owning. Love without possessing.
Teach without making dependent. Lead without trying to control.
Act by helping things happen naturally. Guide by helping things grow
naturally. Empty yourself, and let the Tao fill you and move
through you and use you as part of the pattern.

CHAPTER TWELVE

Joy

Jumping for Joy. On Carrying Vegetables.
Weeding Meditation. Noticing. Simple Pleasures. Sunset.

Jumping for Joy

In a previous era of my life, before the bigger gardens and the duck flock, I kept guinea pigs. I had a garage full of them, about fifty of all ages in a complex naturalistic environment. I spent hours watching their behavior as they interacted, ate, ran around, fought, got along, and made more guinea pigs. Every once in a while I would go to a pet store to look at the exotic creatures. It was painful looking at the guinea pigs, though. They were housed in tiny aquarium-like cages totally open to view. (They hate to be viewable.) Worse yet, they were on top of loose wood shaving bedding so that when they moved their legs, instead of walking or running, their legs mostly just went into the bedding, with little locomotion accomplished beyond a wriggle. These animals had never in their lives run or even walked properly. Racked with pity, I would usually end up buying one.

I would take the guinea pig home and release it on the nice packed dirt in my guinea pig enclosure. The pig would take a few tentative steps, then stop in shock at how effectively it moved forward when it moved its legs. Then the pig would make a short scrambling run, a foot or two. Then, invariably, the pig stopped and performed a spinning leap into the air, complete with a loud exuberant squeal that any fellow mammal could interpret. "Wow! *This* is what my legs are supposed to do! I *knew* something was missing! How wonderful!"

The same thing happened when I bought a baby ferret. It, too, had been in a small aquarium on litter so deep that it could not even walk. It was totally exposed with nothing to tunnel into, nowhere to hide, nothing to explore, and no one to play with—no way to do any of the things that are essential to the nature of a ferret. When I turned the ferret loose on the carpet in the living room, it made a quick dash, then stopped and leaped high doing a full 360-degree spin in the air, then landed and exclaimed excitedly, "Chut chut chut!" ("Hey! This is more like it! So *this* is how my legs are meant to work! I *knew* something was missing! How glorious!") The ferret then ran behind a bookcase and peered out at me. ("Wow! A *tunnel*! I can *hide*. I can zip into this narrow little crevice and hide from that person!") I sat quietly on the rug making no effort to catch the ferret. Next, he zipped out several feet, then zipped backward back into the crevice. Ferrets, we both thus discovered, can run just about as fast backward as forward. What a handy characteristic for a tunnel-dwelling, tunnel-hunting creature! He seemed to use his stiff tail to help guide him in the backward runs. I just watched. He ran out again and explored the entire room in detail, with obvious delight, with intermittent zips forward or backward just for the glory of it and for practice, with occasional jumps into the air complete with spinning and chuttering. Then the ferret came back, approached me, stopped, then jumped from side to side provocatively right in front of me, chuttering emphatically after each jump. ("Let's *play*, Person!") Then he zipped back a bit, then forward, then back and forward again. "Chut chut chut!" he challenged. ("Hey, Person! Catch me if you can!") So we began to play.

I was an air force brat. When I was growing up we moved frequently. My parents didn't garden. I didn't even know anyone who gardened. I spent a good bit of time in fields, forests, and swamps, however, or with my nose in books about nature and science of all sorts. I ended up in molecular genetics. Then, standing in a windowless laboratory one day, I realized something was missing. And I began to garden. When I began to garden, something in me jumped for joy.

On Carrying Vegetables

I once carried a giant kohlrabi around downtown Corvallis for a few hours. It was a 22-pound (10 kg) specimen of 'Kohlrabi Gigante' I had scored in the farmers market. I carried the rab on top of my shoulder, one hand tucked lovingly over it to keep it in place, the stem and leaves projecting behind me as I walked from the market to the library. It was a glorious fall day, sunny, delightfully cool, a perfect day for carrying vegetables. Walking, carrying that rab felt really good.

I thought it was a pretty classically human, normal thing to do . . . hunting and gathering,

you know, which all us primates do—combined with vegetable carrying, which we have been doing at least as long as we have been walking upright, a couple million years or more. Really, nothing new. But from the astonished expressions I garnered, you would think no one in Corvallis had ever seen anyone carry a vegetable before. Okay, it was a large vegetable. And many people have never seen a giant kohlrabi. And I might have also been carrying an outrageously large smile of pride in my possession. Also, I admit that I stood in line at the desk in the library with my rab, and when my turn came I checked the kohlrabi in temporarily so I could amble around and look at books unencumbered. "Guard it with your life!" I told the librarian. "Certainly," she said, not missing a beat. "I'll put it right here on this shelf behind the counter where it will be completely safe. We'll all watch it for you while you find your books." A bit later I checked out my books and my kohlrabi and headed for home, gathering yet more astonished looks the whole way.

Passersby might have been astonished at my rab initially. But then their eyes marked my triumphant grin, and their faces broke into wide grins of their own in response, in delighted, sudden, two-million-year-deep shared understanding.

Weeding Meditation

First thing in the morning in the garden I do any planting that is needed. This takes more disciplined focus and attention to detail than most of the rest of gardening, which takes other kinds of things. Next comes weeding. I do this after planting, but still early in the morning while it is cool and the work is pleasant. Sometimes when I weed I am thinking just about the particular plants I'm weeding and noticing things about the patch I'm working in. Often my thoughts stay focused on that area, but go broader. I am weeding the squash patch that's full of all my breeding projects. My mind floats free and imagines all the varieties these plants might become. Sometimes I get ideas for other breeding projects. Or practical ideas. Hmm. If I plant both this and that, and emasculate these on this day and those on that day, I can do two different breeding projects in this one patch with no isolation. Sometimes what comes are simple observations that I've made hundreds of times, but now I focus on them differently. Hmm. These two summer squash varieties really are the most vigorous. I've grown them each several different years and they always come up first and establish themselves earliest. I should cross the two and see if I can get something even more vigorous. Maybe mass-select explicitly for vigor and see what happens. Sometimes there are random flashes of insight that I will write down later to use in a book.

Sometimes, though, I go into a different state of consciousness entirely while weeding. I feel the gentle breeze on every hair on my arms. The air temperature is so perfect, so soothing, it's hard to tell exactly where my skin stops and the rest of the world starts. The sounds of birds in the adjacent wetland merge with sounds of my hoeing and the sounds of traffic from the highway—wildness, garden, and urban civilization all represented, juxtaposed—myself a part of all three worlds, at home in all of them. I am aware of every muscle as I wield the hoe. Or as the hoe wields me. The person and the garden are one. Bees are working the flowers. A person is hoeing the squash. A caterpillar is chewing a leaf quite noisily.

Where Carol is I'm not quite sure. There is no "I" nor any "Carol" to wonder. Joy is.

Presently, Carol returns, rejuvenated.

Noticing

It's midmorning in the garden. It's getting warmer and I've had enough exercise for the day. I evaluated and recorded the results from a garden trial first thing in the morning. Then I weeded while it was still cool. Then I picked the vegetables for lunch and dinner. Now I wander around the garden just noticing things. This is an open-ended noticing. I'm not looking for anything specific. I'm just looking. Looking in a maximally open frame of mind. Just enjoying being in the garden, wandering around with no goals or purpose, and letting myself see.

There are damp places in the squash patch from the last watering, though the surface of the ground in most of the patch is dry, clearly marking which parts are getting more water than the rest. Are the plants in those spots doing better? If so, it means my squash patch is water-limited and the rest would do better with more water. No. The squash outside the damp areas are doing just as well. All the squash are all getting enough water.

But what's this? This squash plant has a leaf type quite different from the others in the variety. It's bought seed, not a variety of my own. That unusual plant could be an accidental cross. How much variability is there in this material? Hmm. A lot. This entire variety may be crossed up. Don't bother saving seed on this variety this year. Just grow it this year and see if it is what it is supposed to be. I'll know more once it starts making fruits.

In the lower corners of the corn patch the plants are stunted. Not getting enough water. I could do an extra set of the sprinkler, but that would cost labor and most of the water from the extra set would go outside the garden. Instead, when I weed I will thin the corn plants in those corners to much farther apart so they will be productive on the amount of water they are getting. Maybe I'll even eliminate the corn plants in the two dry corners and let the interplanted nasturtiums have all the space. Speaking of which, why are these nasturtium plants so small? They should be tall. That's the only kind I ever buy. But these are a small type. I taste one. Leaves just as tasty. They are actually pretty handy for the dry corners of the corn patch. They don't seem to mind the lower water ration. But the seed I bought was obviously not what it was supposed to be. I don't know what variety these nasturtiums are. The leaves are small compared with those of the tall type; not as good for sandwiches or salads. But the variety is very nice to have in the dry corners of the corn patch. The nasturtiums in those corners are doing just as well as the nasturtiums in the better-watered areas. They clearly don't need as much water as the corn.

The pole bean seedlings emerging in the corn patch are covered with cucumber beetles and are getting eaten up. Obviously this year the cucumber beetles and the pole bean seedlings emerged at the same time. Bad news for the seedlings. However, I planted four or five times as many seeds as I needed just in case, knowing that this planting was near the beetle-problem window. Looks like I'll lose 80 percent of the plants but still end up with a good stand, the beetles having done the thinning. However, the yard-long beans (a kind of cowpea) growing in the same patch in alternate rows are untouched. Interesting. And it isn't a matter of degree. Every common bean seedling has half a dozen or more beetles on it.

Every cowpea seedling has none. And there are a hundred or more seedlings of each. No doubt about it. The beetles don't like the cowpea. I thought they liked about everything—every bean, cucumber, squash. Cowpeas are a *Vigna*, not a *Phaseolus*, so not even the same genus as the common beans, a fairly distant relative. And they are southern-adapted, where insects are a huge problem. Do cowpeas have more defenses against insects than *Phaseolus*? Insects in general or just this one insect? But maybe I'm jumping the gun. Is it cowpeas in general the beetles don't like, or just this particular variety? I have another variety of cowpea interplanted with the corn on the other side of the patch. I'll see when I get over there.

Are any of the common bean varieties less affected by the beetles? Hmmmm. Nope. I have four varieties, dozens of seedlings of each variety, including the heirloom 'Kentucky Wonder' that's good at withstanding things. The beetles seem to be going after and damaging all the varieties about the same.

Let's see, here's the other cowpea variety . . . Yipe! Some of these cowpea seedlings are mottled with yellow. Did they bring a seedborne disease into the patch? Or did they get it here? Well, there isn't the pattern with respect to area of the field or position in the row I would expect if the disease came from the field or if one plant got it here and spread it to the others. There is a random distribution of degrees of disease in seedlings with respect to position in the field. About 20 percent affected heavily, about 20 percent seemingly unaffected, and the rest somewhere in between. The seedlings are just 2 inches (5 cm) tall. And the other cowpea variety is having no trouble. Not a single seedling is showing any sign of yellow mottling. This yellow mottling in one variety is almost sure to be a seedborne disease this cowpea variety has brought in. A virus disease is a good bet. I won't deal with it now, because this calls for serious phytosanitary measures. I'll come back first thing tomorrow with plastic gloves and a plastic bag and pull all the seedlings of this cowpea variety and get them out of here. I'll pull even the cowpea seedlings that aren't showing signs of the disease. At least the ones of this variety. Most or all of them probably also have the disease; even those that don't will have been growing right next to those that do, so might now be infected. I'll keep an eye on the other variety of cowpeas on the other side of the corn patch so if they have become infected, I'll catch it soon and eliminate the entire patch. Had the two varieties been planted close together I would have eliminated both, even the one that looks unaffected. But they are fairly far apart. When I get back home I'll email the seed company and let them know that while I don't care about getting reimbursed, I do want them to know about the situation. No seed company wants to sell anything that carries a seedborne disease.

The diseased cowpea variety doesn't have any cucumber beetles on it either. First approximation is, cucumber beetles don't like cowpeas.

I wander back through the squash patch, then load the tools into my van and prepare to head for home. It was a good gardening day. On a good gardening day there is nothing better. On a good gardening day there is not merely nothing better. There is nothing else.

Simple Pleasures

The first tomatoes of the season are special. My first tomatoes are my 'Stupice', smaller tomatoes but full-flavored. I eat the first of them just plain out of hand. A week later and I have lots more tomatoes. I fix a big salad with just chunks of tomato, feta cheese, a touch of good-quality red balsamic vinegar, and a little pepper. The feta cheese provides the fat and the salt. I taste and get the proportions of tomato, cheese, and vinegar just perfect. That salad is dinner.

Ten days go by and the 'Amish Paste' tomatoes start coming on. Now I have lots of tomatoes, and bigger tomatoes. Now one major meal, lunch or dinner, of almost every day is mostly tomatoes. Cheese, scallions, and gourmet olives often find their way into salads along with the tomatoes. I usually don't mix greens with the tomato salad. Instead, I serve the tomatoes with cheese, scallions, herbs, and dressing on top of a bed of fresh leafy greens or cold cooked eat-all greens. I like my tomato salads on top of the cooked cold eat-all greens best, and make sure to harvest and blanch enough eat-all greens to always have plenty of cold cooked greens in the refrigerator. A chunk of homegrown cornbread on the side provides the calories, always the white neutral-flavored cornbread I make from my 'Cascade Creamcap Flint' corn, which will not clash with or detract from the flavor of the tomatoes as a yellow or more strongly flavored cornbread would.

Then there are tomatoes dropped into soups and stews, just barely warmed, not cooked. I make tomato soup from tomato paste, ketchup, oregano, and red balsamic vinegar, then drop in chunks of tomatoes just before serving.

When the 'Pruden's Purple' tomatoes come on, I make a special celebration for the first meal with these pink-class, 'Brandywine'-type tomatoes. Their flavor is so different from the red tomatoes that they are like an entirely different vegetable, and deserve their own celebration. I serve the first of the pinks just by themselves as the main course for dinner, accompanied by a chunk of cheese and white cornbread.

When the 'Black Krim' tomatoes come on, it's time for spaghetti sauce. I fix meatballs made from grass-fed beef burger. I mix cheese, pepper, and extra oregano into the meatballs before I cook them so they are extra tasty and succulent. The tomato sauce comes from a jar, but I jazz it up with extra oregano, the meatballs, and a little red balsamic vinegar. After the sauce is ready, I drop in chunks of 'Black Krim'. I add so much tomato that I end up with just chunks of warm tomato and meatballs covered with a coating of the spaghetti sauce. I put the sauce-tomatoes-meatballs on top of chunks of white cornbread. Finally, I grate some Romano cheese, an intensely flavorful sheep's-milk cheese, and put a generous amount on top. Magnificent!

Sunset

Gardening brings moments of euphoria. It also brings quieter but longer periods of deep satisfaction. I have finished my gardening for the day. I did exactly the right amount of hard but fulfilling physical labor. Afterward I wandered around the garden looking at everything while munching handfuls of sugar-podded peas. Then I gathered the food for dinner. This day I did some work, noticed something new, realized something, used my brain and curiosity to make myself a better future gardener. Now I am pleasantly, perfectly tired. I am relaxing in a lawn chair overlooking the garden, swigging water occasionally, gazing out over the adjacent wetlands to the mountains beyond outlined against the blaze of red and orange of the setting sun. I'm happy in this garden, tending this rich, generous fertile soil, nurturing these plants. This is who I am. This is who I am meant to be. This is what I am meant to do. This is where I belong.

A squash, kale, and pea jungle.

This patch of 'Alexanders Greens' has been volunteering for six years and providing tasty greens from midwinter to late spring each year. It is growing in almost full shade on solid clay and has never been fertilized, watered, or weeded.

When lambsquarters came up as the main weed in this section of the squash patch, I left the plants until they were at this prime size for harvest. Then I plucked off the top of each plant and hoed down the remains.

I like to pick tomatoes into mono-layers in cardboard trays to prevent their bruising in transporting and storing. These 'Stupice' tomatoes are very early and have great flavor.

A nice batch of 'Black Krim' tomatoes with one red tomato, a 'Cosmonaut Volkov', in the upper right for color comparison.

Garden Woman Meets Pigweed with Attitude

Garden Woman: "Tra la la la la. A pole bean picking I will go . . . Yipe! What are *you* doing here!?!" A giant pigweed has sprawled over the whole area. Garden Woman cannot even get to her beans.

Giant Pigweed: "Ha ha ha! Puny human! You ignored me when I was little. Now my stalks are half an inch across and solid wood at the base. You'll never get me out of here now! You can kiss your beans good-bye!"

Garden Woman: "Humph! We'll see about that!" She reaches for her secret weapon, her heavy-duty peasant hoe.

Garden Woman assumes her special secret grip with both palms facing downward, her stand such that the hoe is working off to her side, not directly in front of her. Her arm muscles work to lift the hoe, but then simply guide it as it drops, letting gravity do all the work. She orients the hoe so that one of the sharpened flared points enters the plants or ground first, creating a slicing action backed by the full force of the heavy blade.

She simply lifts and drops, using arm muscles and swaying a bit with legs and hips, her back comfortably straight the whole time. The hoe slices through the woody stems and roots of the pigweed like butter.

Garden Woman: "Take that, Evil Pigweed!" (Thump! Thump! Thump!)

Giant Pigweed: "Ack! Ack! Ack!" In just a few minutes the pigweed is history.

Garden Woman: "Hmm. These onions could use a little weeding, too." It's just ordinary weeds in the onions—weeds with succulent, not woody, roots and stems. But the spacing between the plants is tight. So Garden Woman turns to her Coleman hoe.

The Coleman hoe requires its own secret grip. Garden Woman grasps the hoe handle like a broom, with both thumbs up; the razor-like blade skims just beneath and almost parallel to the surface of the soil. Garden Woman's back stays comfortably straight while she uses the hoe. In a few short minutes the onion patch is weed-free.

In about fifteen minutes with the right tools Garden Woman clears all the weeds from the beans, the surrounding paths, and the onion patch. Then she turns to the pigweed piled in the paths.

Garden Woman: "I know you would just re-root if I left you in the garden, Foul Pigweed. So I'm going to put you in this tub until you are all dried up. Die Pigweed, Die!"

Giant Pigweed: "AARRrrrggggggggg."

Garden Woman: "Well, it may not look as neat as Eliot Coleman's gardens, but at least I can pick my beans."

Farmer Paul Harcombe thins our seed crop planting of 'Gaucho' bush dry beans using a Coleman hoe.

Nick Estens uses a Glaser wheel hoe with a furrower attachment to make planting furrows for our squash seed crop.

I planted this bed of eat-all 'Green Wave' mustard in mid-March in space set aside for tomatoes and harvested it eight weeks later when it was time to transplant the tomatoes. I harvested a measured 18 pounds (8.2 kg) of prime greens from the 4-by-16-foot (1.2-by-4.9-meter) bed for eating and freezing for winter.

The perfect polyculture: A did-it-mostly-themselves patch of squash and 'Russian Hunger Gap' kale.

Leaf radishes are a very fast, versatile eat-all green. The variety shown here is 'Saisai Hybrid'.

'Indian Spinach—Red Aztec Huauzontle' (*Chenopodium berlandiera*), a fast-growing summer eat-all green.

'Green Calaloo' is the fastest-growing amaranth I've found so far. It's an excellent warm-weather eat-all green.

Eat-all greens are excellent as catch crops. There were gaps in this corn planting. Instead of wasting the space, I filled it by sprinkling and raking in a little 'Green Calaloo' amaranth seed. These little eat-all patches are now ready to harvest.

'All-Red' amaranth grows a little too slowly to make a good eat-all crop, but its color is spectacular. If you like amaranth in salads, it's worth growing just for that.

Summer eat-all variety trials: three beds planted simultaneously (July 2). The first bed is all amaranth varieties. Left to right: 'Emerald Isle', 'Asia Red', 'Green Pointed Leaf', 'All Red', 'Green Round Leaf', 'Red Stripe Leaf', 'Green Calaloo', 'Redleaf' (behind the 'Green Calaloo' and not shown; it barely came up). 'Red Stripe Leaf' and 'Redleaf' (which is actually striped, and might be the same as 'Red Stripe Leaf') seem to be the most widely available leaf amaranths, but they consistently grow miserably in my trials. The 'Green Calaloo', the only variety of this batch I consider suitable as an eat-all green, is way past optimal harvest stage. Notice the redundant information in marking trials. Each variety is labeled at the left corner. But also, varieties are arranged so that those adjacent to each other are different enough to be visually distinct. In addition, I have field notes that record every variety in order.

Summer eat-all variety trials, bed 2. Left to right: 'Suehlihung #2' mustard, 'Purple Orach', 'Green Wave' mustard, 'Indian Spinach—Red Aztec Huauzontle'. The real winner from this July 2 planting is the 'Indian Spinach'. The 'Green Wave' and the 'Suehlihung #2' mustards are workable, but do much better in cooler-weather plantings. Both were also a bit too crowded to grow optimally, so I'm harvesting them at this stage. The 'Suehlihung #2' and 'Green Wave' grow equally fast, counting only height, but the 'Suehlihung #2' produces little biomass compared with the 'Green Wave'. The 'Purple Orach' doesn't grow fast enough to make a good eat-all; no orach I have tested does. Notice how the label between the 'Purple Orach' and 'Green Wave' mustard has mysteriously vanished, but it is still easy to tell which is which, because the two green mustards are separated by the purple orach.

In summer eat-all trial bed 3 I alternated five different gai lohn (Chinese kale) varieties with four different radish varieties. The gai lohns all grew at about the same rate and looked about the same, as did the radish varieties. However, the radishes all quickly outgrew and overgrew the gai lohns. The gai lohns all work fine as eat-alls planted by themselves.

This 5-foot-square (1.5-meter-square) eat-all patch of 'Saisai' leaf radish was planted July 4 and harvested August 14, forty-one days later. Average plant density was 3 inches (7.6 cm) apart in all directions. The plants were never thinned or weeded. Plants grew up to 2 feet (0.6 m) tall. I harvested by cutting off the entire patch at 6 inches (15.2 cm) above the ground with a serrated bread knife, leaving behind everything that wasn't prime in the process. When harvesting, I keep the plants oriented with the stems in one direction so kitchen prep time is minimized.

This one small 25-square-foot (1.5-square-meter) patch of eat-all leaf radish yielded a measured 13.3 pounds (6 kg) of fresh all-edible harvest for a yield of 0.53 pound per square foot (4 kg per square meter) of garden. I cooked the three flats of greens in three batches and filled fourteen freezer containers with 2½ cups (15 ounces) each of blanched, drained greens, each topped off and covered with the cooking water—the amount I like for a single (generous) serving of eat-all greens. Then I ate one portion, put a couple away in the refrigerator for the next two days, and froze the rest for winter use.

I like to plant small successional patches of peas instead of one big patch. A commercial tomato support is an easy way to support a small patch of medium-vine-length peas. These 'Oregon Giant Sugar' edible-podded peas are at the prime picking stage for maximum flavor.

道

*Square without corners, vessel without walls, soundless tone,
formless image—that is the shape of the Tao. Accept, allow, adapt, flex,
bend, change—that is the method of the Tao. Create, complete, fulfill,
reverse, return, renew—that is the movement of the Tao.*

*Every particular thing is a manifestation of the Tao. The Tao gives rise
to all things. The Tao fulfills them. To the Tao they return.
That is why everything venerates the Tao. And that is why we
find the Tao when we look inside ourselves.*

*There is a time for living and a time for dying, a time for
planting and a time for reaping, a time for motion and a
time for stillness, a time for working and a time for rest.*

*The Tao gives birth to all things, nourishes them,
shelters them, cares for them, comforts them,
and in the end, receives them back into itself.*

道

CHAPTER THIRTEEN

Completion—Seeds

Cycles and Circles. The Do-It-Yourself Seed Bank. You Will Not Fall Off the Edge of the Earth If You Don't Save All Your Own Seed. Preparing Seed for Long-Term Storage. Containers for Storing Seed. Eight Seed-Saving Myths. Creating Your Own Modern Landraces. Rejuvenating Heirloom Varieties. Breeding Crops for Organic Systems. Dehybridizing Hybrids—Disease-Resistant Tomatoes. Tomato Genes and Genetics. Breeding the Heirloom Tomatoes of Tomorrow.

Cycles and Circles

Fall 2003. I sit by my mother's bed, a hospital bed set up in the living room. She no longer knows I am her daughter, or even that she has children at all. I'm a comforting, reliable presence she can depend on. That's all. I tell her stories about the garden and the crops. She can't remember the stories, but she enjoys hearing them. We hold hands. Her vision is poor and speech is difficult. But she understands. She signals emotional intensity by squeezing my hand. The crops are all in now, I tell her. The squash are tucked indoors safe from any coming freeze. The corn is stacked in piles in front of fans in the downstairs. I've been scurrying around getting everything taken care of before the first frost. Now I can relax. My mother smiles. "That's . . . good," she says. She squeezes my hand.

A few days later my friend Blanche and I sit next to Mother's bed, me holding my mother's hand, Blanche holding mine. Mother is unconscious now. I can "feel" her still in there, though. Her every breath is a struggle. I ask for help. I reach out somehow, and out loud I say, "I'm asking whatever there is out there to help with this transition." Three seconds pass. Suddenly I feel an immense elation, an incredible joy. And I do not feel my mother's presence anymore. I turned to Blanche. "She's gone," I say. Blanche nods. The body takes one more laborious breath. Thirty seconds pass. Another breath. Then the pulse in the neck slows, fades . . . ceases. It was the end, the completion.

Fall 2007. Nate France and I are doing some projects together on a farm he is living on. Then his world falls apart. His marriage breaks up and his farm situation evaporates. I have space in the downstairs of my house. So Nate moves in down there, and we continue gardening together on some land I lease nearby while he both recovers and then . . . falls in love. Sarah and little two-year-old Ava move in with Nate.

Fall 2010. Nate, Sarah, and Ava go off in new directions of their own, just coming back to visit occasionally.

Fall 2013. I'm writing a new book, *The Tao of Vegetable Gardening*. Nate and Sarah come back for a visit, this time bringing both Ava and new baby Gwendelyn.

In some areas of the country everything in the garden dies or becomes dormant in winter and is covered with a blanket of snow. Here in western Oregon we have occasional snows. But the garden has plenty of overwintering greens. Yet the cycle of the garden is just as definite as in places with more rigorous winters. All my main storage crops—potatoes, corn, beans, squash—must be harvested in late summer or fall before the rainy season or the first freeze. After October I will do no more planting or garden tending until February or March. It is too cold, wet, and muddy. Winter in the garden is a time of rest. That does not mean the gardener is resting, though. She isn't. Or at least, she does so only briefly. The summer annual plants have died, but they made seed. My garden activities aren't over; they simply move indoors. Every spare bit of space in the house is covered with tarps full of dry plants or piles of corn, beans, or squash. It's both the end and the beginning, another stage in the cycle, another part of the circle. It's time to tend to the seed.

The Do-It-Yourself Seed Bank

Seeds tucked away in a vault in Svalbard, Norway, are fine as far as they go. But in hard times this is unlikely to be very far from Svalbard, Norway. Even in the best of times you, an individual gardener or farmer, won't be able to get any seeds out of the Svalbard seed bank. Only institutions that have deposited seed can get seed out, and then only the seed that they deposited in the first place.

A generic can of "survival" seeds is also not likely to be very useful. Such cans are full of varieties that are not optimal for your region, nor best for your purposes. Nor is the can likely to include more than trivial amounts of the big-seeded staple crop varieties such as corn, legumes, and squash,

the very crops that would be most important in the event of any major disaster.

The best seed bank is the one that is full of vigorous, regionally adapted varieties of exactly the crops you care about the most, your very favorite varieties, those that do best for you, that you already know how to grow and use. It contains serious amounts of seeds of these crops, enough to plant a normal-sized crop of each of them for three years or more, not just a tiny sample of seed that has to be increased for several years before there is enough to be useful. And the best seed bank is in your own home or neighborhood.

Every gardener should have her own seed bank. Even if you aren't a seed saver, you should have your own seed bank. Even if you never experience any disaster beyond the ups and downs of ordinary living, it's useful to have your own seed bank.

First, if you buy seed, you can buy in bigger amounts and pay just a fraction of what you pay when you buy a little packet every year. You simply dry the bought seed for long-term storage, divide it up into one-year-sized packets, and put it into your freezer.

Second, there are years when we can be too poor or too busy to buy seed. With my own stash of seed, I buy seed when I can afford the money and time and dip into my stash when I can't. Some years I buy a lot of seed. Some years I buy none, but still plant everything I want. The personal seed bank gives me greater flexibility and resilience in good times as well as bad.

Third, many varieties these days are produced by only one grower in the country or world, even if sold by many retail seed companies. So you can find your favorite variety suddenly stamped with CROP FAILURE in every seed catalog one year.

When that happens it's very nice to have your own stash of the variety.

Finally, the commercial seed trade regularly loses varieties. And the fact that a variety is widespread and very popular is no guarantee. Of my five favorite squash varieties from the 1980s the commercial trade has lost all of them. Often many seed companies continue selling the crossed-up trash under the traditional name. The more reputable seed companies drop the variety when it is no longer what it is supposed to be. The result is that you can buy varieties sold under the classic name long after the real variety that name represents no longer exists anywhere. All that has actually been preserved is the name.

One reason often given for saving our own seed is to be able to prevent the loss of varieties we care about. Where the seed isn't available otherwise, we have to do our own seed saving, of course. But when the seed is widely available we just need to stash some away so that, if it is lost, we have good seed we can use to start our own seed saving after it becomes necessary. I really wish I had just dried and frozen a little of every variety I liked thirty years ago. If I had, I would still have real 'Sugar Pie' pumpkins and real 'Guatemala Blue Banana' winter squash. (What is being sold by these names today bears little resemblance to and has little of the distinctive flavors they had thirty years ago.) And I would not have had to do all that work to breed the 'Sweet Meat—Oregon Homestead' line of winter squash after the commercial trade lost 'Sweet Meat'.

Even if you purchase most or all of your seed, only varieties that can be saved belong in your seed bank. The seed bank needs to feature open-pollinated varieties. If you cache hybrid seed you have only the seed you have stored and are at a dead end when it is used up. Whether your seed bank starts with saved or purchased seeds, its usefulness depends on your being able to save seed and perpetuate the variety after taking it out of the bank. So the seed in the seed bank needs to be of open-pollinated, not hybrid varieties. Hybrid varieties don't breed true. Some hybrids don't produce seed at all. Others do, but the seed produces plants with variable characteristics that reflect segregation for all the genes and traits that were different in the two parent varieties that went into the hybrid. Some hybrids can be dehybridized into uniform, open-pollinated, true-breeding varieties that resemble the original hybrid, but this is a breeding project that usually takes a number of years. And not all hybrids can be dehybridized. The more your gardening uses open-pollinated rather than hybrid varieties, the more complete your seed bank can be, and the more seed-supply resilience you have.

If you regularly use some hybrid varieties, search for equally good open-pollinated varieties you can switch to. Most hybrids in most crops are not superior to the best equivalent open-pollinated varieties. They are simply marketed more intensely by the company that produces them than are varieties for which no one has a monopoly. Furthermore, with open-pollinated varieties, the fact that gardeners and farmers can save their own seed puts a limit on what can be charged for it. Hybrid seed makes a captive market of anyone who becomes dependent on it. We gardeners have no reason to encourage and promote monopolies in our seed supply. So we need to do our homework and trial more varieties than just the heavily touted ones. We need to actively search for and replace the hybrid varieties in our gardens with varieties that are open-pollinated.

You Will Not Fall Off the Edge of the Earth If You Don't Save All Your Own Seed

For some crops and varieties you may not be able to save seed nearly as good in quality as what you can buy. It depends on the crop and where you live. Seed saving is fun but is also hard work, and it takes up space indoors for piles of drying and dried plants waiting to be threshed. Where seed of a variety that I like and use is available in as good or better quality than what I can produce myself at prices I can afford, I'm happy to buy the seed.

I save my own corn, beans, and squash seed, because I breed these, and because I can produce far better quality of seed for corn, beans, and squash than anything I can buy. I also save seed of the rare varieties that are available in only small amounts at expensive prices; you can't produce very big crops of these without saving your own seed. But I don't save seed of everything, not even of all my very favorite varieties.

I don't produce my own seed of 'Oregon Giant Sugar' pea, for example. We organic growers in the Willamette Valley have not yet figured out how to grow pea seed without getting big weevils in nearly every seed. (The weevil is *Bruchus pisorum*, the green pea and dry pea weevil.) Freezing kills the insects. But a big dead beetle that takes up a quarter of the volume of almost every seed kills many seeds and damages most of the rest. The weevil eggs are actually laid on the immature green seedpod; the larvae burrow through the pod and into the developing seed and grow into adult beetles by the time the seed has finished developing and drying

down. So freezing the dry seed kills the already fully developed beetles rather than preventing their development. So I buy wholesale amounts of OGS pea produced in places that don't have the beetle. If the variety is ever lost commercially I'll do my own pea seed saving starting with my stash, and will just have to live with the poorer seed quality. But unless or until then I'll buy the seed.

The maritime Northwest is an ideal place to produce brassica seed and seed of overwintering biennial varieties such as beets. We can overwinter the plants and produce fat, heavy seed on huge plants. When the seed is produced elsewhere it is usually tiny compared with seed grown here. If I lived anywhere else I would do just enough brassica and winter biennial seed saving to know how, but would buy my seed of those crops from the Northwest rather than growing it myself. (Unless I wanted to breed my own varieties or adapt varieties better to local conditions, of course, which would require saving my own seed.)

In addition, specialization is legitimate. I don't save all my own brassica seed even though I live in the ideal place in the world to grow it. I buy a lot of it from other maritime Northwest growers and seed companies who are already doing a good job of producing it. I don't have enough room in my house for big piles of dry plants of all the brassicas I grow waiting to be threshed. My house is too full of big piles of corn, beans, and squash.

Beware AAS varieties. This acronym stands for "All American Selections." It could more

accurately be said to stand for "All Agribusiness Selections." The rules of the contest require winners to pay a certain percentage of all world-wide seed crop sales for a number of years to the AAS organization. You can do this only if you totally control the seed, which you can do only with hybrid, patented, or PVP (plant variety protected) seed. So AAS is an award for the best new *proprietary* varieties, that is, the best new hybrid, PVP, or patented varieties only, not the best new varieties. Nor are these best new proprietary varieties better than preexisting public domain varieties. Nor are the "new" characteristics for which they are touted necessarily new. Sometimes they have been around in public domain varieties for years or decades. When you see the words *AAS Winner* be aware that the variety is proprietary, and that you cannot grow it without giving up your traditional ability or right to save your own seeds and some of your seed-supply resilience.

PVP varieties are open-pollinated varieties and breed true to type. However, they have legal restrictions that forbid you from selling, swapping, or even giving away seed. You are, though, allowed to produce seed for your own use or sell a vegetable crop grown from such seed. It is also legal to use PVP varieties to do crosses to develop new varieties of your own. (It isn't legal to derive a new variety simply by selection from a PVP, or to use a PVP to make a hybrid whose seed you sell.) So it is workable, though not ideal, to include PVP varieties in your seed bank. In a mega-disaster there probably wouldn't be anyone around to enforce the PVP rules. But most uses of seed bank seed will be in situations short of mega-disaster. And for seed-savvy gardeners and farmers, selling or distributing seed is an important option. I suggest avoiding all PVP varieties in your garden and seed bank unless the variety is genuinely superior to all other equivalents.

A third category to avoid putting into our seed banks is varieties that are sold only with seed treatments (fungicides). The seed treatments involve dangerous chemicals that are not practical to apply on the home scale. The varieties are not resistant to common diseases that are found in most gardens, so cannot grow without the treatments. The combination of a variety not resistant enough to ordinary diseases to grow without seed treatments and the seed treatments themselves makes for varieties for which you cannot save seed. Avoid buying treated seed. Avoid varieties that are sold only as treated seed.

Of course, we don't want GMO varieties in our seed bank. Even if we didn't mind the genetic modifications, GMO varieties usually have utility patents that are maximally restrictive. You are forbidden from saving seed even for your own use or for using the variety in breeding of any kind. You basically don't own the seed when you buy it. You are just leasing the use of that amount of seed for one year.

In most cases your seed bank will start as a small box in a freezer. It may grow to a whole freezer of its own if you get ambitious and figure on storing enough seed for years, or for your entire neighborhood as insurance against a serious disaster. Properly dried for long-term storage and put into a freezer, the seed will keep virtually indefinitely as long as the power lasts, and for several years beyond should the power stop. During that several years beyond, you would save and replace the seed as you used it.

If at some point you want to expand your do-it-yourself seed bank to a whole freezer, consider getting a used chest freezer. These are often available free because people get rid of them to convert to more convenient upright freezers. Chest freezers are fine for seed, as you don't need frequent or convenient access. Most used chest freezers change hands via freebie ads placed by those wanting to get rid of or acquire one.

A seed bank can be designed around the needs of ordinary gardening or it can be oriented toward getting your family or neighborhood through serious hard times or even mega-disasters. I think the best do-it-yourself seed bank does at least some of both.

A seed bank oriented toward ordinary gardening need only contain the varieties you usually use in amounts good enough to cover, say, three years of gardens of the size and type you plant in ordinary times. A survival seed bank is designed at least in part to deal with the possibility of serious hard times or even mega-disasters—situations in which the family, neighborhood, or region might suddenly become much more dependent on the food it can grow itself. The survival seed bank can and probably does include the ordinary-gardening seed bank. But it needs to be bigger, and it should have a much larger proportion of seeds of staple crops.

The survival seed bank must be capable of providing seed for more garden(s) than you have during ordinary times. You may need to greatly expand the size of your garden in serious hard times. You may need to provide seed for many friends or for an entire neighborhood, not just yourself. You and others in your neighborhood might join together to farm a large field, which you are able to do in part because you can provide the seed. Or you may swap starting seed to others for a payment in the resulting crop. The seed itself might be a more useful reservoir of value

and medium of exchange than any of the paper or plastic currently used for the purpose, so you might want extra seed specifically for trading. The survival seed bank should include enough seed to give you many options.

The survival seed bank also needs to focus primarily on staple crops, crops that provide large amounts of storable carbohydrates and proteins, such as corn, beans, and winter squash. If you actually needed to survive on your garden produce, crops that produce serious amounts of carbohydrates and proteins are critical, and the long-storing grains and grain legumes are the first priority. Corn is easier to produce on a small scale with hand tools than are wheat and other small grains, and will be the logical grain choice for most gardeners. Dry beans and dry (soup) peas are also easy to grow. Next in priority are crops that provide serious carbohydrates and that can be stored at least for some months, such as winter squash, turnips, rutabagas, onions, and beets. (Potatoes are invaluable too, but if your seed bank includes potatoes it must be in the form of true seed from seed balls—produced by the flowers—not tuber "seed potatoes," and such true seed of potatoes, alas, does not breed true. But true seed will still make potatoes of some sort, which is likely to be better than no potatoes of any sort in hard times.)

If you have a good-sized garden, you can experiment with growing different varieties of these survival crops on a small scale. (There are major chapters on growing, processing, and using corn, beans, squash, and potatoes in my book *The Resilient Gardener: Food Production and Self-Reliance in Uncertain Times.*) You may actually produce a good fraction of your staples. Or you may produce just enough to learn how, to have enough for the occasional batch of spectacularly delicious polenta or cornbread, and to figure out which varieties to stockpile in the survival portion of your seed bank.

If you have too small a garden to experiment with staples, your main task is to gain some experience so you are not trying to grow your very first batch of cornbread corn or harvest and thresh your very first batch of beans in hard times. The experience does not need to be on your own land. You can simply volunteer to help with the planting, harvesting, and handling on a bigger garden or small farm that grows the kinds of crops you want to learn about. Many people volunteer to help produce seed crops for my own seed company as well as those of other private breeders of public domain varieties. They help support the development and distribution of public domain varieties while getting a share of the crops and the practical experience growing them. Many CSAs encourage volunteers. And if you are a student, consider working on an organic farm that produces the relevant staple crops for one or more summers.

The survival seed bank has to have serious amounts of grains and pulses in it, enough to plant gardens with major amounts of these crops in really hard times. Many gardeners don't grow field corn or dry beans, so aren't familiar with the amounts of planting seed they would need, the amounts of land needed, or even the size of the crop they would want or the space needed to store it. Here are a few useful numbers.

A standard 5-gallon (18.9 liter) pail holds roughly enough shelled corn to provide all the calories for a month for a person on a 2,000-calorie-per-day diet. I worked out the numbers for 'Gaucho', a bean that is one of my staples, and got that same 5-gallon figure.

You would not want a diet of just corn and beans, of course. You would want to grow and eat plenty of other things too. And you may need more than 2,000 calories per day, especially if you are doing a lot of physical labor. The one pail per month is just a reference number for calculations. Suppose, for example, you want to provide 1,000 calories of corn and 500 calories of beans per day per person for a family of four for a year (with the rest from other things). You would need one-half pail of corn per month per person and one-quarter pail of beans per month per person—that is, six pails of corn and three pails of beans per person for the year. For a family of four that would be twenty-four pails of corn and twelve pails of beans. I think this would be a reasonable goal if you needed to live primarily from your gardening. I would be figuring on a 3,000-calorie-per-day diet because of the serious physical gardening and other labor that hard times would probably require, with half the calories from grain and pulses and the other half from other high-carbohydrate garden produce such as fruit, winter squash, potatoes, sweet potatoes, beets, turnips, and onions. And of course we would want to continue growing all the other not-so-caloric garden produce such as greens for their vitamins and minerals.

Pails filled with grain can be stacked four high without crushing one another or falling over. If you moved a bed far enough away from the wall in a bedroom and stacked a single four-pail-high rank of pails twelve pails long between the bed and the wall, you could fit forty-eight pails in the space.

How much land would it take to grow a pail of corn or beans? My 'Cascade Ruby-Gold Flint' corn is an early corn that can produce ears by mid-August in a region with little summer heat and under conditions of only modest fertility. I bred it to be the ultimate survival corn variety as well as to make the most delicious cornbread and polenta. CRG makes solid-colored ears of several different colors ranging from deep red, red, and red-brown to maple-gold, gold, and yellow; each color gives a different flavor of cornbread. In my far-from-ideal corn-growing climate CRG yields about 3 gallons per 100 row feet (37 liters per 100 row meters). So it takes about 166 row feet (51 m) of garden to produce a pail of corn.

If you grow a full-season corn in a place with more summer heat (such as in the Corn Belt), your corn might yield up to about twice as much as mine, even sticking to traditional open-pollinated varieties that taste good and using hand-gardening methods. (Hybrid varieties can yield even more, but much of the higher yield depends on soluble fertilizers and on the tighter spacing you can get only with specialized tractors and herbicides. And none of the hybrids taste very good as people food; they are bred for maximum yield for animal feed. Nor does the seed breed true.)

How much seed do you need to plant to grow a pail of corn? I like to plant my CRG corn at 4 inches (10 cm) apart in the row, then thin plants to every 8 to 12 inches (20 to 30 cm). At that spacing, it takes 300 seeds per 100 feet of row (980 seeds per 100 meters of row). CRG seed is fairly big at 1,200 seeds per pound (2,650 seeds per kilogram). So a 1-pound bag of seeds plants 400 feet (122 meters) of row, enough to grow 2.4 pails of corn. (A 1-kilogram bag of seeds would plant 270 meters of row, which would give you 5.3 pails of corn.) And to plant the space it takes me to grow a pail of corn requires 0.42 pound (6.7 ounces/191 g) of corn seed.

(Note: The CRG corn and most of the squash seed I sell through Fertile Valley Seeds is already dried and packaged for long-term storage and can just be tossed in a freezer as is for an instant survival seed bank. And I sell CRG in 1-pound survival packs as well as smaller packets.)

The 'Gaucho' bean gives me about 12 pounds per 100 row feet (17.9 kg per 100 row meters). This is a bit more than half the corn yield. If I call it half the corn yield, it means I need twice the land to grow a pail of beans as I do a pail of corn. So I figure it takes me about 332 row feet of garden (100 m) to grow a pail of 'Gaucho' beans.

When it comes to using the seed in your seed bank and expanding it after it is no longer otherwise available, you will need to know how to save seed. However, you don't need to know how to save seed to start a seed bank. You should start your seed bank right now, whatever your seed-saving experience or lack thereof. If worse comes to worst you might have to learn seed saving quickly and are likely to make mistakes initially. But that still beats knowing all about seed saving but having no seed.

If you don't know the perfect varieties to stockpile, just start with modest amounts of the varieties that are recommended by gardeners in your region. Then add bigger amounts of your favorites as you have a chance to check out varieties and develop your own preferences.

You can start with just one variety and a small jar in a corner of your freezer. Is there some variety you would never want to be without? Start there. Doing something, however modest the scale, is better than doing nothing. My own seed bank is a few varieties of corn, beans, and squash plus eat-all greens varieties and assorted other odds and ends. At the moment it's fewer than twenty varieties. I'm working my way toward, probably, about fifty varieties. Twenty to fifty varieties isn't everything. But one can eat very well on twenty to fifty carefully chosen, beloved, familiar varieties.

All you need to start creating your own seed bank is a little space in the corner of your freezer and one set of skills. You need to know how to dry seed for long-term storage, the subject of the next section.

Preparing Seed for Long-Term Storage

Seed as it comes from the average seed company in paper packets or bags may seem dry, but it is actually fairly moist and is respiring significantly. If you seal such seed in an impermeable plastic bag or jar it will suffocate. Such seed should also not be frozen because it is wet enough that much or most of it will be killed if frozen. So to prepare seed for long-term storage we need to dry it additionally, then seal it into airtight containers. Then we freeze the containers. However, even if stored at room temperature, such dry seed goes into a deeper dormancy than ordinary moist seed, and will have greater longevity than ordinary seed. I routinely get nearly 100 percent germination on five-year-old corn seed in both germination tests and under field conditions, for example, when I have simply grown the seed optimally, dried it properly, sealed it in appropriate bags or in glass jars, and left it at room temperature.

Unless you live in a desert, getting seed dry enough for long-term storage usually requires artificial indoor drying. Most seed savers use electric dehydrators. Putting the seed near a wood-burning stove or other source of heat or spreading it on screens in front of a small electric heater-fan unit should also work.

The two major commercial styles of dehydrators used for drying seed are the little round-tray dehydrators (many makes and models) and the Excalibur dehydrators. The ideal temperature to dry seeds is 95°F/35°C. Don't bother getting a dehydrator that doesn't have a thermostat or that has one that doesn't go as low as 95°F/35°C. The little round-tray dehydrators don't hold very much, don't have much power, and the holes in the middle of the trays make unloading the trays a nuisance. However, these dehydrators only cost about $50 new, and are often available in yard sales for almost nothing.

Up until recently I've been a big fan of Excalibur dehydrators. These have a much better design and hold much more than the little round-tray dehydrators. Unfortunately, the company has changed to a thermostat that does not go down to 95°F/35°C in reality. It actually runs at 20°F/11°C or more hotter. A friend who bought an Excalibur in spring of 2014 was able to get the company to send him one of the older, much better thermostats. The Excaliburs, depending on size, cost about $200 to $350. They are still unexcelled for drying fruit and vegetables, but alas, with the new inferior thermostat, are no longer suitable for drying seeds. However, you might call the company and see if you can still get one with the older thermostat. Otherwise, I would suggest building something of your own when you get beyond the little-round-dehydrator scale.

Whether you buy a dehydrator or make one, or use a less formal drying situation, don't trust it until you have checked it out with a thermometer.

I usually do my seed drying in winter after the heat is turned on and the air and seeds start out drier than it is in summer. I dry most seed for a day, then test the dryness. For flint corn, dent corn, or beans, the test is the "hammer test." You put a seed on a brick or on the concrete driveway outdoors and whap it with a hammer. If the seed shatters, it is dry enough to seal into airtight containers and to freeze. If the seed smashes or mushes it is still too wet. Flour corns smash instead of shattering, however dry they are. So just use the same drying time as with the flint and dent corns. With squash seed, peel off the shell and bend the seed. If it snaps the seed is dry enough. If it bends or is mushy when it breaks it isn't. (If you just bend the whole seed it can fool you, as with big squash seeds the shells might be dry while the meats inside still are not.) Small seeds I mostly just dry similarly to the large seeds. There's a bit of guesswork to it.

I freeze-cycle all planting-grade corn seed that will be stored more than a year, even if I don't plan to store it in the freezer. Corn seed usually has eggs of seed storage insects in it. Freezing the seed for at least a few days kills the eggs and insects. Some people need to freeze-cycle their bean seed. It depends on which types of grain storage insects you have in your region.

Containers for Storing Seed

To keep seed dry you need to store it in airtight containers such as glass jars or heavy-duty plastic bags of the correct type, or plastic or metal buckets. Otherwise the seed simply reabsorbs moisture from the air. Glass jars are impervious to insects and rodents but not to earthquakes. Plastic bags are sometimes burrowed into by insects and are no protection against rodents. So far I have not had insects or rodents get into seed in sealed, airtight, heavy-duty plastic pails. When the seed is stored in a freezer, the freezer itself is a big metal can that protects its contents from insects and rodents . . . even if the electricity has long been off. So I use mostly heavy-duty plastic bags for freezer-stored seed.

I store food corn and beans in 5-gallon (18.9 liter) plastic pails at room temperature. (I don't run the seed through the dehydrator process or freeze-cycle it if it is for food or if I plan to plant it within a year or two.) For intermediate-term storage of seed corn I also use the pails, but I dehydrator-dry and freeze-cycle the corn in plastic bags to kill insects, then put the bags in the pails. The pails are then kept at room temperature.

For plastic bags suitable for storing seed as well as for 5-gallon (18.9 liter) plastic pails I suggest uline.com (1-800-295-5510). They have by far the best prices I have seen for containers of all kinds, take orders by phone or online, and routinely deliver on the next business day. The Uline catalog has so many kinds of bags you could spend all day trying to find the right ones, however, so I'll be specific. The plastic bags I use for long-term storing or freezing seed are the 4-mil reclosable polyethylene bags. I buy the S-1707, S-1302, and S-1712 to cover sizes of 3 × 5, 4 × 6, and 6 × 9 inches, respectively (7.6 × 12.7 cm, 10.2 × 15.2 cm, and 15.2 × 22.9 cm). (They cost $35, $41, and $67 per carton of one thousand bags, respectively.)

The pails are 5-gallon (18.9 liter) heavy-duty high-density polyethylene plastic pails, which are food-grade (though the description fails to mention that). They are S-7914 ($4.69 each) with the S-9948 lids ($1.40 each). Make sure you get one of the pail-opening tools; it makes life lots easier.

Eight Seed-Saving Myths

The do-it-yourself seed bank is based on the premise that, if you need to, you can save your own seed. If you want to grow rare varieties, you usually buy a small, expensive packet that contains just enough seeds in it to give you a start. You have to be able to save your own seed of the variety thereafter. In many cases you can produce seed that is much better than what you can buy. Where seed is expensive or your garden large, saving your own seed can reduce your expenses considerably. But above all, saving your own seed is necessary if you want to adapt varieties to your own needs, climate, and growing conditions or create new varieties of your own. Crop varieties incorporate the values of their breeders. When we save seeds and create our own varieties, we are saving and propagating our own values along with the seeds and varieties.

There are four basic aspects to seed saving. The first is just physically saving the seed. That's the easy part. Plants are eager to produce seed. Where the seed is dry, such as kale or cabbage seed, we generally just cut the dry scapes or whole dry plants and pile them on a tarp and thresh them out by stomping on them. Then we clean the seed by winnowing in front of a fan (or outdoors on a day with a gentle breeze). For most wet seed like squash seed, we just wash it and dry it. For some seed, such as tomato seed, we do a fermentation step to dissolve and remove the gelatinous layer clinging to the seeds. Then we wash and dry seed.

The second aspect of seed saving is assuring that the seed is actually genetically pure seed of the variety. That's harder. Plants mostly don't care about pure varieties. Virtually all varieties within a species will cross readily with all other varieties in that species. So to produce pure seed of a variety we usually need to isolate it by some recommended distance—the isolation distance—from other varieties of the same species. (Alternatively we can use various other tricks to achieve purity such as hand pollinating with squash—pollinating with the pollen we want and taping the flowers shut to exclude undesired pollen.) So when you start your seed saving, you start off not by collecting seed from some plants. You start by planting in such a way that the variety you want to save pure seed of will be workably isolated from other varieties of the same species.

The third aspect of seed saving is numbers. We need to save seed from enough plants to avoid inbreeding depression, that is, loss of vigor associated with inbreeding. So we don't usually save seed from just one corn plant. We want a minimum of a hundred or more. (Corn is very sensitive to inbreeding depression.) In addition, in some cases we are trying to preserve genetic heterogeneity, such as in a landrace or open-pollinated variety. The more genetic heterogeneity we have in a variety and the more we want to preserve the heterogeneity, the greater the number of plants we should grow each generation.

Finally, we need to continue selecting the variety. Varieties deteriorate spontaneously as a result of mutations and contaminating crosses. To maintain a good variety we must always rogue out off-types each generation and save seed only from the plants that are true to type for the variety.

Any good book on seed saving will give you information on not just physically saving the seed but also the isolation distances and numbers of plants recommended for each crop species. As a first book on the subject, as well as an essential reference for seed savers at all levels,

I recommend *Seed to Seed* by Suzanne Ashworth. *Seed to Seed* will tell you how to save the seed of just about every garden crop you can imagine. (Get the second edition, which also gives you basic growing information on all these crops.) The information in *Seed to Seed* is solid, and the isolation distances recommended are real rather than gross underestimates based on wishful thinking. (Many books and seed catalogs sometimes give wishful-thinking isolation distances for certain crops. In my experience, where Ashworth differs from other books Ashworth is invariably more realistic and reliable.) So if you are going to buy one book to keep with your do-it-yourself seed bank, I would recommend *Seed to Seed*. It's the Seed Saving 101 of the field.

Seed to Seed only goes so far, however. In many cases it tells you that you need to isolate different varieties of the same species by half a mile (about a kilometer), for example, and leaves you sitting there with your garden right next to those of neighbors with no way to save seed of the crop other than to move to somewhere lots more isolated or do laborious caging. And, like most other seed-saving books, it covers little on how to actually do selection.

The seed saving information that constitutes part 2 of my book *Breed Your Own Vegetable Varieties: The Gardener's and Farmer's Guide to Plant Breeding and Seed Saving* is intended to be Seed Saving 201. It picks up where *Seed to Seed* leaves off and covers such subjects as how rigorous your methods need to be in various situations, how to most effectively cheat at isolation distances and plant numbers, and how to select effectively for what characteristics you want and avoid selecting accidentally for those that you don't. I also cover seed saving for corn, beans, and squash in detail in the respective chapters on those subjects in *The Resilient Gardener*. In this book I'll limit my further discussion of seed saving to a few myths that come up repeatedly among seed savers at various levels.

Myth #1. You don't have to read any books about seed saving in order to do it because Indians didn't read any books and they did it.

Indians had very sophisticated cultural traditions for how they grew and handled crops that included pragmatic methods for saving seed and keeping varieties pure. The book *Buffalo Bird Woman's Garden* describes the agricultural patterns of the Hidatsu and Mandan Indians of the upper Midwest in enough detail that we can see how they kept their varieties pure. Buffalo Bird Woman planted different varieties in separate blocks, interspersing blocks of bean varieties with blocks of corn varieties to give some physical isolation to both. She planted five different varieties of corn, but each had a different maturity time and distinctive seed type. The different maturity times helped provide temporal isolation for each corn variety so crosses would be less likely to occur. And the distinctive seed types made it easier to identify accidental crosses when they did occur and eliminate them. Buffalo Bird Woman also made arrangements with neighboring gardeners so that they planted the same variety where their gardens were adjacent.

Buffalo Bird Woman did not know about the role of pollen in plant reproduction. She used methods that had been determined to work empirically. "Corn traveled," her people believed. That is what they called it when they saw the signs of crossing between corn varieties. The

Hidatsu and Mandan Indians deliberately practiced cultural methods they had learned that minimized corn "traveling." They knew that the closer two varieties were planted, the more the corn traveled between them. They knew that corn traveled more readily over open land than broken land. (Trees and bushes break up wind patterns so pollen spreads less readily.) Buffalo Bird Woman learned how to keep her varieties pure as part of the agricultural traditions of her culture. In addition, she grew large enough blocks of each variety that most of the seeds she saved from the crop would be from the interior of the block instead of the edges. Seeds from the interior of a sizable block of one variety are least likely to be crossed with other varieties. Many European or pioneer American farmers grew only one variety of many crops, and grew it on a large enough scale so that it tended to be self-isolated.

Myth #2. You won't get crosses if you use the recommended isolation distances for various crops.

Sorry. No. Generally, the farther apart two varieties are the less they cross, but it's a matter of degree, not absolutes. You will usually still get some crosses at the recommended distances. How much isolation distance you need with any crop depends not just on the tendency of the crop to outcross but also on how large your planting is, how large the plantings of possible sources of contamination are, how important a few outcrosses would be, how pure the variety is as it stands, wind patterns, pollinator pressure, and other factors. In addition there are many tricks that can be used to isolate crops other than distance. And in many cases being able to recognize and eliminate crosses is easier and more practical than avoiding them. (I discuss all these

aspects in detail in my first book, *Breed Your Own Vegetable Varieties*.)

Myth #3. If you have a number of varieties that might cross and are doing seed saving, it's best to plant the varieties that look similar near each other.

No. This standard practice doesn't lower the rate of contaminating crosses but instead makes it more difficult or impossible to recognize and eliminate unwanted crosses when they do occur. You are most likely to lose varieties when there have been crosses that you initially cannot recognize.

It's better to plant the varieties that are most different next to each other. If you plant a bean variety with purple plants next to one with green plants, for example, hybrids will usually be pink. Next generation, you simply eliminate all the pink plants before they start to flower. That lets you keep the varieties pure even if there have been crosses. In addition, you will actually learn something about how many crosses you get in your garden when you plant two varieties of that species side by side.

Myth #4. When you save seeds to maintain a variety, you just keep all the seed.

No. You usually keep seed from only the best plants. You must actively eliminate the worst plants in every generation. If you don't, the variety will deteriorate spontaneously. There is actually no such thing as "maintaining" a variety. There are always new mutations and accidental crosses. You have to actively select to maintain a variety. (Where a variety is genetically messy as a result of prior neglect or accidental crosses, you may keep seed from only the best half the plants, for example. Where a variety is genetically pretty clean you may need to eliminate only a few plants among hundreds or thousands.)

Myth #5. It's obvious what the best plants are.
Nope. Sometimes it is. Often it isn't. "Best" depends on what the variety is supposed to be, not some absolute. If you just save the biggest fruits you are often selecting for smaller numbers of fruits, for example. If you select the biggest ears of corn you are likely to select for lateness if you aren't careful, as bigger, later plants often have bigger ears. It takes deep thought about what the essence of the variety is and what your own needs and wishes are and how those two aspects might intersect to figure out how to select when saving seed of any given variety. I usually don't save seed the first year I grow a variety. I have to become familiar with the variety first.

Myth #6. Common beans (Phaseolus vulgaris) are inbreeders and don't cross. So you don't have to isolate different bean varieties from each other.
Sorry. No. It's true, beans *are* inbreeders, but nearly all so-called inbreeders still do cross. They simply tend not to do it as frequently as the plants we choose to call outbreeders. How much beans cross depends on your region and other conditions. If I plant two rows of beans of different varieties adjacent to each other I will get more than 5 percent crosses. That's enough to lose both varieties completely in a few generations unless I also take measures to recognize and eliminate the crosses.

Myth #7. You can tell when you have crosses on seed like corn or beans. The crossed seed looks different.
Sometimes, and only sometimes, this is true for corn. False for beans. With corn, crosses of black-seeded and yellow-seeded varieties sometimes do

show on the seed. Black × yellow kernel color will usually (but not always) show up as bluish or speckled kernels on the yellow ear, but only sometimes show up well enough to be apparent on the black ears. Crosses of sweet and field corn are always obvious on the sweet ears but not on the field ears. Crosses of flint corn and flour corn show up on the flint but not usually on the flour. It depends on which genes are dominant. Crosses involving pericarp (skin) color on the kernels don't show up because the mother plant makes the pericarps. The genes in the seed don't have anything to do with it.

You can't tell crosses at all on bean seed. With beans, the skin (hence bean color) is made by the mother plant, not the seed. So the beans reflect the genes in the mother plant. Likewise, the size and shape of bean seed is determined largely by the pod, which is also made by the mother plant. So if there are crosses between two bean varieties, you don't see any sign of it until you grow out the resulting seed.

Myth #8. You should always isolate a given crop by the specified recommended isolation distance and grow the specified recommended number of plants.
Not necessarily. Sometimes you can afford to be pretty casual. It depends on your role and purpose. If you are selling or distributing seed you should be rigorous with your isolation distances and numbers. If you are the only source for the seed, if the variety's entire existence depends on you, you need to be more even more rigorous. If you are saving seed just for yourself, however, and have some backup seed in the freezer, you can be pretty casual about everything.

For the squash seed I sell, for example, we plant only one variety of each species in a farm field that is separate from my home garden. Each variety is isolated by miles from others of the same species. And there are a hundred or more plants of each variety to keep the heterogeneity and vigor up. However, I'm the breeder for these varieties and the ultimate source for the seed.

For squash varieties where I am saving seed for just my own use and have backup frozen seed, I am quite casual. I grow several varieties of each species in my home garden. Isolation distances? Hey, I just put a variety as far from other varieties of the same species as possible in my home garden and leave it at that. I may simply alternate varieties of different squash species to provide a little isolation within each block of each species. Numbers of plants? I grow whatever number I was growing anyway just for food. If the variety gets crossed up or overly inbred I can go back to my frozen stash for a fresh start. If I improve the variety, I add the improved version to my seed bank. One of the advantages of having a do-it-yourself seed bank is that it allows you to do more seed saving with less labor and fuss.

Creating Your Own Modern Landraces

Traditionally, a landrace is a locally adapted variety of domesticated plant or animal that has usually been created largely by mass selection rather than by inbreeding in family lines. Landraces are usually more phenotypically as well as genetically variable than most modern pure varieties. That is, the individual plants are usually obviously visibly different from plant to plant in addition to being different for invisible characteristics at the gene level. The individual plants within a given landrace are uniform for certain critical characteristics but variable for many others. In most cases the obvious phenotypic variability comes along with additional invisible genetic variability, both of which contribute vigor and resilience to the landrace. In some cases the variability is desirable. Variability in maturity times that spreads out the harvest can be desirable when the family is eating the food. If some plants are more drought-resistant and some are more heat-tolerant, and so on, the landrace may more reliably produce food in different seasons with different weather than would a more genetically uniform variety. Above all, traditional landraces are selected to perform well and meet the needs of the people who developed them.

Sometimes we do not actually need pure varieties. I think our modern breeding has gone too far in the direction of worshipping uniformity for its own sake, that is, for over-inbreeding and creating uniformity for many characteristics beyond those that matter. Inbreeding eliminates genetic heterogeneity. And genetic heterogeneity is usually associated with more vigor and productivity.

In many cases modern home gardeners want several different colors of leaf lettuce or beets, and actually plant multiple varieties or even buy a blend that is a mix of several pure varieties. The gardener is artificially creating some of the advantages of a landrace by using such mixtures. But she might be better served by growing a highly genetically heterogeneous landrace that gives multiple colors instead of the mix of pure varieties. The heterogeneous landrace would likely be much more vigorous than the pure varieties. And it is easier to save the seed of a single landrace than to save the seed of several pure varieties.

If you pretty much like all colors and shapes of summer squash, for example, instead of growing pure varieties you might let several varieties you like all cross with one another, and start saving squash from the mix each year. What you get when you plant a seed from this material will be pretty unpredictable. But as long as only summer squash went into the mix, it's likely to produce good summer squash. And if you simply start selecting for earliness or productivity or whatever you care about, pretty soon you are likely to have your own landrace that performs better under your conditions than did any of the pure varieties.

Alan Kapuler frequently interplants two or more varieties and allows them to cross, then selects for vigor and production in subsequent generations under his conditions, but allows variability in plant form or colors. He calls such interbreeding mixes of offspring segregating out of crossed populations "grexes." They are one kind of landrace. For example, he produced 'Three Beet Grex' starting with three heirloom beet varieties: 'Yellow Intermediate Mangel', 'Crosby Egyptian Purple', and 'Lutz Overwintering'. 'Three Beet Grex' produces beets of gold, Day-Glo orange, orange-pink, or red. It's much more vigorous than any of the pure gold-beet varieties. And it is apparently widely adapted. It has gotten rave reviews for being the most vigorous beet in New England. And a friend of mine in Missouri says it does so well for him that it has become one of his major staple crops both for people and for animal feed. ('Three Beet Grex' is available from Peace Seeds, Peace Seedlings, or Fedco.)

When I got crosses between my dry bean 'Gaucho', a gold Argentine heirloom bean (*Phaseolus vulgaris*), and 'Black Mitla', a supposed Mexican tepary bean (*P. acutifolius*), the crosses segregated out black, brown, tan, and gold beans and were *extremely* productive. And the mix of

beans has a delicious, intensely beefy flavor. So I just continued growing and using the mixed-up material, which, after a few generations, I named 'Beefy Resilient Grex'. I just select for a bit more earliness and for yield. (The selection for yield happens automatically if you keep nearly all the beans.) 'Beefy Resilient Grex' is much more productive than both parent varieties and has a huge amount of genetic heterogeneity, which helps the plants deal with disease or weather problems.

When Joseph Lofthouse interplanted every *Cucurbita moschata* squash variety he could get his hands on, initially none matured completely on his Utah mountain farm. (See chapter 9.) He simply interplanted all the varieties and let them cross up. After a few generations of growing his squash, he ended up with 'Lofthouse Landrace Moschata', which gives squash of all sizes and shapes, but they are all early.

I love the earliness, vigor, and size of the 'Lofthouse Landrace Moschata', but I would prefer all or mostly all butternut shapes. That shape better fits my use patterns. So I am going to select for just butternuts. I could do this by self-pollinating a few plants for a few generations to produce pure varieties. But if I did that, I would be eliminating much of the genetic heterogeneity and, at the same time, probably much of the material's vigor and productivity. Instead I plan to just gently mass-select for the butternut shape under open-pollinating conditions. That is, I will grow out dozens of plants and just save seed from the butternut-shaped ones without doing any controlled crossing, inbreeding, or hand pollination. In a few generations I should have nearly all butternut shapes, but the material will still retain all the rest of the genetic heterogeneity and, hopefully, its vigor and productivity. That is, I want to reshape the 'Lofthouse Landrace

Moschata' to better fit my use patterns, but am happy to leave it as a landrace.

Landraces are not optimal for all purposes. I'm often too fussy about flavor or specific uses to find landraces appropriate. Many people are happy with landrace open-pollinated corn, for example, because they use their corn mostly to make cornbread, and you can make wet-batter cornbread with any field corn. And they mix their cornmeal with wheat flour, which dilutes its flavor, making the flavor of the corn less important. However, I need my flint corns to be very pure for flint type, because only very pure flint corn makes good quick-cooking delicious polenta from whole-ground corn with no special screening to remove floury components. And I breed flint corns for their flavor as polenta as well as cornbread. Likewise, I want my flour corn varieties to be very pure flour corn types so that they can be used to make fine-grained cakes and cookies and gravies, not just cornbread. And I use all my varieties to make cornbread with 100 percent cornmeal and no wheat. So the flavor of the corn is critical. However, I'm willing to let my corn varieties contain multiple colors—but only those colors that vary by the whole ear rather than kernel by kernel so I can sort whole ears to get different flavors from one variety. And I only allow certain colors within my varieties, those I've figured out are associated with great flavor.

I also wouldn't be interested in growing a landrace summer squash, because I am fussy about summer squash flavor, and most of the plants of a mixed pool of intercrossing squash would be unlikely to give me flavors as good as those of the pure varieties I like. In addition, I like to grow varieties that are tastier than most as fresh squash but that are also delicious when dried to be a winter staple. That way all the squash that escape prime summer squash size can be dried in summer to make soups and stews in winter. (For more on this see *The Resilient Gardener.*) However, most varieties of summer squash are tasteless when dried. In an intercrossing population derived from many varieties, very few plants would produce fruit that tasted good dried, and there would be no way to tell which these were without laboriously drying and tasting fruit from every plant. So I will stick with pure summer squash varieties.

When I bred the 'Hannan Popbean' garbanzo, I started with USDA material that was a Middle Eastern landrace uniform for seed color, flavor, and the popping characteristic. But there was variability for everything else. Plant maturity ranged from early July to just starting to flower in October after my rainy season has started and freezes are threatening. This meant that the land was tied up from March until late fall and that many plants were too late to yield at all with the amount of summer heat I have. And only the earlier plants could finish without irrigation, so the entire plot had to be irrigated. The number of seeds per pod varied from one to four, so seed size varied from tiny to much bigger. This is okay if the beans are parched in hot sand, but I wanted to roast or microwave-pop them, which requires uniform seed size. Some plants were tiny. Some were big but crept along the ground and were no more than a few inches tall. Others were very erect and 3 feet (1 m) high. And there was everything in between. The landrace seemed to be uniform for resistance to soilborne diseases but variable for resistance to aphid-transmitted legume diseases, for which the Willamette Valley is a hotbed. However, that huge genetic variability included resistance to all the aphid-borne diseases. At least in some plants. Many plants died of disease when small or became obviously diseased and had to be culled, however.

I needed bigger, more uniform seeds for the parching methods I was using. In addition, I needed the plants to be resistant to my local diseases. And part of what I envisioned was a legume that could be grown without irrigation in my no-summer-rains region. I had too many requirements for a variable landrace, even a landrace adapted to my land. I needed a pure variety. So I mass-selected for plants that had one or two big seeds per pod, that were disease-free under my conditions, and that could be planted in early March and would finish drying down their seed in late July. My 'Hannan Popbean' grows well without irrigation and is uniformly resistant to pretty much everything. But at this point it is a pure variety. There is little or no obvious physical variability from plant to plant within the variety. (However, since I mass-selected instead of using inbreeding, there might well still be plenty of invisible genetic variability.)

Here are four useful guidelines for creating your own landrace. First, when you start a landrace by crossing or by interplanting two or more varieties so they can cross, do not start selecting in the first generation. The characteristics of the first generation have mostly to do with which genes are dominant. You can't see the characteristics associated with recessive genes until the second or subsequent generations. So don't start selecting until the F_2, the second generation after the cross.

Second, in most cases it's best to select gently. You don't have to eliminate all of the plants with some undesired characteristic in any given generation. In fact, in most cases you probably shouldn't. If you don't like something, just save less seed from ones like that. In the early generations you usually don't have all possible traits in all possible combinations. Not all possible traits even show up in the first few generations. So don't eliminate too much too fast. In fact, if you have the space, it's perfectly reasonable to do no selecting at all for the first few generations and just let all the genes and characteristics get more thoroughly mixed together.

Third, it's usually good enough to save seed based on just one parent. So when I select based on squash characteristics, I'm selecting based only on the female parent. But I only need gentle selection to create a landrace (or to maintain an established variety). Selection based on just one parent is usually good enough.

Finally, be open-minded and opportunistic. Taste and cook whatever shows up in various ways. Start separate lines where appropriate. You might pool all the seed from the plants that produce early but smaller fruits, and all the seed from the plants that produce bigger later fruits, for example. You might end up with two or more landraces that serve different needs from one project. Just eat the fruits from the plants where the two landraces are adjacent. And a few crosses between the landraces doesn't hurt. All you really need is a little semi-isolation, not true isolation.

Rejuvenating Heirloom Varieties

Sometimes we obtain a much-touted famous heirloom or other open-pollinated variety and plant it eagerly only to find that the plants are so wimpy and lacking in vigor that it is hard to grow them at all. And they produce so poorly that they aren't worth the space or time. If it is a winter squash, for example, the plants may produce fewer and smaller fruits than they should. The flesh might be thin. And the fruits might mature late or not at all. And there can be weird characteristics, such as defects in flower form. In many cases what has happened is that the variety has become genetically bottlenecked.

Genetic bottlenecking is a result of saving seed from too few plants. We might start with a variety that has enough invisible or visible genetic heterogeneity to be vigorous, but if we save seed from just one or a few plants it's as if all the alleles (forms) of all the genes have to pass through a bottleneck to get to the next generation, and many of the alleles don't make it.

Genetic bottlenecking has two bad effects. The first involves deleterious mutations and alleles. New mutations are always occurring in the plants of any variety, and these are often somewhat deleterious. That is, they can make the plant less vigorous or lower yielding under some or all conditions. Or they can make the food lower in quality. If the mutation is dominant and the effects are large, visible, and obvious, we usually cull the plant carrying the deleterious gene or it culls itself by being unable to grow or thrive. But often the mutant gene is recessive, that is, it affects the plant only when it is present in two doses. In addition, many detrimental genes have only small effects or are detrimental only under some conditions. If we grow just a small number of plants we have a harder time identifying good plants. And if we save seed from just a single plant we have a much higher likelihood of the planting being homozygous for one or more deleterious mutations so that our line of the variety will then be fixed for that mutation from then on. So, for example, where d is a recessive deleterious mutant allele and D is the normal version of the gene, we might have started with a line where most plants were genotype DD but some were Dd and there was an occasional dd. And when we saved seed from just one plant we might have, alas, chosen a plant that was dd. If so, from then on, all plants in the line will be dd. The line has become fixed, that is, pure-breeding for the deleterious mutation.

The other problem with genetic bottlenecking is that it removes invisible genetic heterogeneity, and heterogeneity itself often confers vigor and resilience. So when a variety has been genetically bottlenecked, it can be wimpy and unproductive because it has either lost essential genetic heterogeneity or become fixed for deleterious genes or both.

With famous heirloom varieties, often there are many independently saved and transmitted lines of the variety available. And sometimes when we obtain a number of lines, one will turn out to still be vigorous and productive enough to be worth growing. But often all the lines have been bottlenecked to some extent. However, if the lines have been independently maintained, we can often recombine the lines and regenerate a vigorous, productive line that is everything the variety is supposed to be. And it takes relatively minimal effort.

Just obtain as many lines from as many independent sources as possible. (Ideally, you get lines from seed savers or small seed companies that have maintained their line themselves, or can tell you where they got their line from initially so you can avoid redundancies.) Then you interplant all the lines and proceed exactly as I described in the prior section for creating landraces. You just start by interplanting different lines of one variety instead of different varieties. You let all the lines open-pollinate and intercross in all possible combinations. You deliberately interplant in such a way as to get all possible combinations of crosses. Then you save all the seed without selecting that first generation and plant it out. You will usually have some much more vigorous plants the very next generation. However, in most cases you should simply save seed from all the plants again. Then the following year (the F_2 and beyond), you just start mass-selecting for the characteristics the variety is supposed to have as well as for vigor and productivity. Often you can produce material that is mostly vigorous and productive and worth growing in just a few years. Most of the problems of the lines you started with vanish.

The reason why this works is that, while most of the lines might have lost much or most of their genetic heterogeneity, each line has lost different components. In addition, each line has picked up at least some new genetic heterogeneity through new mutations and accidental crosses. So by adding the genes of several lines back together we can often restore or re-create enough genetic heterogeneity to make a vigorous, productive variety again. Alternatively, where the problem is that lines have picked up deleterious mutations, different lines have often picked up different deleterious mutations. So when the lines are allowed to cross and recombine, there are actually good versions of all genes present in the mix that can sort out into a new line that is not fixed for any deleterious mutations.

When I started breeding 'Cascade Ruby-Gold Flint' corn I started by interplanting 'Abenaki' (aka 'Roy's Calais') and 'Byron'. Both are eight-row New England flint corns, but they have enough different characteristics so that it was obvious they had been maintained separately for a good while. Each variety had some characteristics I wanted. 'Abenaki' has big ears for an eight-row corn and many are red, which makes cornbread with a unique and spectacularly delicious flavor. But I lost most ears of 'Abenaki' because of poor husk coverage that promoted bird damage. (A bird-damaged ear in my regions gets colonized by aphids, so the whole ear turns into aphids and aphid muck and becomes inedible.) In addition, the yellow ears of 'Abenaki' were pale, uninspiring yellow, and the interior color of the kernels was light yellow. So I chose to interplant with and cross to 'Byron', which has good husk coverage, deeper interior gold color, and a maple pericarp color. 'Byron' also had pencil cobs, however, which I did not want. (They break up or don't shell out as well in my hand-crank corn sheller.) Plus, 'Byron' had smaller ears and kernels than 'Abenaki', and the yield wasn't very good. It also did not have the red pericarp color, which gives the corn its unique flavor in cornbread. When you want something new with some of the characteristics of one variety and some of the characteristics of another, you start by just crossing the two varieties. Both varieties were very pure flint types.

But in addition, I figured that most Native American and pioneer corns have been genetically bottlenecked. If both varieties were bottlenecked, if I simply intercrossed them and kept the numbers I saved seed from high enough to preserve the genetic heterogeneity created by the cross, I should be able to get material that was much more productive than either parent. This turned out to be true. 'Cascade Ruby-Gold Flint', an early corn, is not only more productive than its parents but also as productive or more productive than the most productive full-season open-pollinated flint corn, 'Longfellow'. ('Longfellow' is probably also bottlenecked, so is probably not living up to its full-season potential.) When I crossed the two different varieties I created a lot of genetic heterogeneity. Then I saved seed from several hundred plants per generation thereafter so as to maintain as much of that heterogeneity as possible while I mass-selected for the characteristics I cared about.

Breeding Crops for Organic Systems

When we organic gardeners and farmers create our own landraces, breed new varieties of our own, or adapt varieties as we save their seed, we are growing and selecting under organic growing conditions. Exactly what these conditions are varies wildly from garden to garden and farm to farm. In some cases it might mean that the plants must be able to thrive on a regimen of modest soil fertility. But not necessarily. Many organic gardens and farms have very rich soil. Whatever else, however, crops that excel under organic growing conditions need to resist many insects and diseases without the aid of insecticides or fungicides. In addition, good organic varieties usually need to be able to compete effectively with weeds.

The first method for obtaining good organic varieties suitable to your specific conditions is to simply do all your plant breeding and selection and seed saving under your growing conditions. You will automatically be selecting for varieties that do well under organic conditions, and under exactly your version of organic growing conditions.

Sometimes a good organic variety thrives because it has specific genes that confer resistance to specific insects or diseases. But I think in many cases the good organic variety thrives not because it has a specific gene with which to meet every specific problem, but because it is just so vigorous that it outgrows most problems. The good organic variety may have resistance to a particular insect, for example. But in most cases it doesn't. It gets eaten by the insect just like everyone else. But the variety grows so fast that it shrugs off the insect attack as minor damage and produces excellent crops anyway. And nothing helps a variety compete with weeds better than just growing fast and being very vigorous.

So a second method I use for developing good organic varieties and adapting varieties to organic systems is to simply focus on rapid growth and plant vigor. I select very strongly for fast germination, rapid growth, and vigor in the seedlings. This has the delightful advantage that is one of the easiest things to evaluate and select for. In addition, selecting for it can be done with virtually no extra land or labor. All that is required is careful attention to how much seed we plant and how we thin.

So whether I am developing a new variety or just saving seed with an established variety, to get powerful selection for growth rate and vigor, I start by planting at least three times as many seeds as plants I need. Then I thin to the desired spacing, *but only after the seedlings are big enough to have had time to express their own genes* rather than just coasting on the food and biochemistry the mother plant puts into the seed. How long this takes I have come to understand through noticing and considering seedlings that carry spontaneous mutations that prevent the formation of chlorophyll in peas and in corn. Such mutations are lethal. The seedlings that carry them are completely yellow. However, a pea or corn seedling that can't make chlorophyll grows just as fast as the rest of the seedlings until it is about 2 inches (5 cm) high. Then it comes to an abrupt halt and grows no further, then fades, shrivels, and dies. What this tells us is that up until the pea or corn seedling is about 2 inches (5 cm) high, its growth and vigor depend almost exclusively on the stored food and biochemistry synthesized by the mother plant. Only after the seedling is bigger than 2 inches (5 cm) is the contribution of the mother plant diluted enough so that the seedling's performance expresses its own genes rather than those of the mother. So I let seedlings get as big as possible before I thin so that they have had as long as possible to display the virtues or liabilities of their own genes. With corn, for example, I let the seedlings get at least 4 inches (10 cm) high before I thin them. I leave them to grow even bigger if there is room. I also want to avoid having the plants crowd one another too much and make thinning too difficult, however.

Another reason for not thinning earlier is that more shallowly planted seed often comes up first. If we thin too early we can be thinning based on accidents of planting depth instead of based on genetic capabilities of the seedlings. By just waiting a bit, all the seed has a chance to emerge, and the minor differences in emergence rate associated with accidents in planting depth are overridden by the vigor and growth capability of the individual seedlings.

Most seeds are smaller than pea or corn seed and don't carry as large a store of food and biochemistry from the mother plant. So crops with smaller seeds may be expressing their own genes earlier than at 2 inches (5 cm) high. The same general principle applies, however. Sow excess seed and

thin as late as practical in order to select for rapid germination and growth and plant vigor.

Selection requires eliminating plants, choosing. If we sow about as many seeds as the plants we need and keep all the plants, we aren't selecting. If instead we plant excess seed and thin optimally we get very powerful selection for germination rate, growth rate, and general plant vigor in every generation, and our entire stand of plants is already selected heavily for vigor before we even get around to choosing which plants to save seed from based on other characteristics.

A corollary: It's often possible to tell which varieties will make vigorous organic varieties just by noticing how fast they germinate and how rapidly the young plants grow.

A third method I use in developing organic varieties is to focus on genetic heterogeneity. Some varieties are highly inbred and have little genetic heterogeneity but are nevertheless quite vigorous. This is most common in crops such as peas, beans, or tomatoes that are basically inbreeders and are not very sensitive to inbreeding depression. These varieties have been selected for having genes that allow vigorous plants even when present in homozygous combinations. But many varieties and crops derive some of their vigor from genetic heterogeneity itself. When a plant or variety has a lot of genetic heterogeneity this means it has many genes in configurations like *Aa* or *Bb* or *Cc* rather than *AA*, *aa*, *BB*, *bb*, *CC*, or *cc*. That is, the plant is heterozygous rather than homozygous at many genetic loci (positions). (Why genetic heterogeneity tends to confer vigor is not yet understood.)

Some genetic heterogeneity is visible; it affects the way the plants look. But much or most genetic heterogeneity is invisible. So a vigorous heirloom variety might seem pretty uniform, but may actually have a lot of invisible genetic heterogeneity. Greater genetic heterogeneity is often associated not with just greater vigor but also earliness, yield, and general resilience of a variety.

We increase genetic heterogeneity when we cross different varieties. We start with two varieties that are *AABBCCDD* and *aabbccdd*, respectively, and get an F_1 hybrid that is *AaBbCcDd*, for example.

We decrease genetic heterogeneity drastically when we inbreed. For every generation of inbreeding we lose half the genetic heterogeneity. You can see this by considering a single gene that is heterogeneous in a given plant, *Aa*. If the plant is self-pollinated, that is, inbred, the cross with respect to this gene is *Aa* × *Aa*. Our high school genetics tells us the offspring will be ¼ *AA* + ½ *Aa* + ¼ *aa*. Now notice that one-fourth of the offspring are now homozygous *A* and one-fourth of the offspring are now homozygous for *a*. Both *AA* and *aa* have lost all their genetic heterogeneity at this locus. Only half the offspring are heterogeneous with respect to this gene. So we started with one plant that was genetically heterogeneous at the locus, that is, 100 percent of the plants were genetically heterogeneous. And we wound up with only half the offspring being genetically heterogeneous with respect to the gene. The same thing applies to all other genes that start off heterozygous. So we lose half the genetic variability with a single generation of inbreeding. Should these offspring be inbred another generation, an additional half the remaining genetic variability will be lost. This loss in genetic heterogeneity can mean lost vigor and resilience.

After we do a cross to start a breeding project we often then have to do some inbreeding to create varieties that are uniform for the characteristics we care about. Most pure-breeding squash varieties are inbred four or more generations, for example. I think one way we can breed more vigorous organic varieties is to limit the inbreeding to what is absolutely essential by not requiring uniformity for characteristics that don't matter. These days squash varieties are uniform not just for squash flavor, size, shape, and color, for example, but also for leaf color and shape and many other characteristics that don't matter. When we require a squash variety to be uniform for leaf size, color, and form we do so only by discarding large amounts of genetic heterogeneity unnecessarily.

Another way we can maintain more genetic heterogeneity as we breed varieties is to mass-select (as I described in the section on creating landraces) rather than inbreeding at all. When we mass-select instead of doing familial inbreeding, however, we often end up with a variety that isn't totally uniform for some of the characteristics that do matter. We should simply accept this. In fact, this pattern is characteristic of good heirlooms. They are vigorous and resilient but have a few percentage of off-types. This is probably because they were created by mass selection instead of familial inbreeding. They contain recessive genes that, when they appear in homozygous combinations, give rise to off-types. We simply rogue out the off-types each generation so that the frequency of the genes that cause them drops, or at least becomes no larger in the next generation.

When we mass-select for a characteristic associated with a recessive gene it's pretty easy to fix that gene, that is, get it homozygous in all plants in the variety. But when we mass-select for a characteristic associated with a dominant gene there are almost always a few recessive genes hanging around in the variety invisibly that occasionally give rise to off-types. The few off-types each generation is a small cost to pay for the greater vigor and resilience we can get in many varieties by avoiding too much inbreeding.

A final method I use is to go ahead and do familial inbreeding when I'm creating a variety, but then restore the lost genetic heterogeneity afterward. I inbreed and create not just one but several independent lines, then intercross them toward the end of the project as I described doing in breeding 'Candystick Dessert Delicata' in chapter 9. Each inbred line will have lost much of its genetic heterogeneity during the familial inbreeding phase. But different lines will have randomly lost different components of the initial heterogeneity. By crossing the lines at the end of the project I can restore most of the genetic heterogeneity but still have uniformity for all the characteristics I selected for when developing the lines.

Dehybridizing Hybrids— Disease-Resistant Tomatoes

As I described in the chapter on tomatoes, the newer nastier strains of late blight are making it increasingly difficult to grow blight-sensitive varieties of tomatoes outdoors in most of North America and elsewhere. We are going to need blight-resistant tomatoes. Many university and seed company breeders have been working to find genes that confer late blight resistance in wild tomato species, introgress them into domestic varieties, and release them in new blight-resistant varieties. However, much of what is being offered is in the form of hybrids. Very expensive hybrids. A single seed of a late blight resistant tomato may

cost you 50 cents, for example. So we may want to dehybridize those blight-resistant hybrids. In addition, we usually often start a breeding project by crossing two varieties, that is, making a hybrid of our own. Then the rest of the project involves dehybridizing that hybrid into a pure-breeding, open-pollinated variety that has all the characteristics we care about. If we want delicious heirloom-style tomato flavors in the future, we will need a new generation of heirloom tomato varieties that incorporate serious resistance to late blight. We will need to produce these varieties ourselves starting from commercial hybrids or our own hybrids. And we will need to know how to breed for disease resistance.

Most modern tomato hybrids carry many dominant genes that confer resistance to a number of different diseases. For example, the hybrid might be produced by crossing a variety that is resistant to *Fusarium* wilt, nematodes, and early blight to a variety that is resistant to late blight, tobacco mosaic virus, and *Verticillium* wilt. Since all those resistances are associated with dominant genes, the hybrid will be resistant to all six problems. Many of the genes confer resistance to only certain strains of a disease, so sometimes resistances to different lines of a disease are stacked into the hybrid. In addition, the two parents usually differ for many other characteristics, so when we save seed of a hybrid the seed will give rise to a next generation of plants that will show segregation not just for disease resistance, but for all other characteristics that were different between the two original parent lines. So we might see segregation for plant type, fruit size, shape, color, earliness, or other characteristics in addition to disease resistance. This section will focus on late blight resistance, but to dehybridize for resistance to other diseases (or to nematodes), the same process applies. All these resistances bred into the commercial hybrids are associated with dominant genes. And in many or most cases the hybrid is heterozygous for the resistant genes.

Selection for disease resistance can be pretty straightforward if your land or region is so full of the disease that the sensitive plants invariably and obviously become diseased every year. However, in many cases we don't have any problem with the disease at all some years. And often some parts of a field are more affected than others, and plants next to heavily diseased plants get more heavily exposed. So even when the disease is present we can't necessarily always tell the difference between resistant plants and plants that were just lucky and didn't get exposed as much or as early as most of the rest. University breeders often inoculate plants with spores of the disease or have test fields full of the disease. But we gardeners and farmers don't normally have such trial fields. Let's take the worst-case scenario, the most difficult breeding situation possible—let's imagine that most years we don't have any problem with the disease at all, then one year out of every few the weather conditions or other factors are just right for the disease, and it kills every sensitive plant in our garden. In this case, we can't identify resistant plants most years. However, we need not let these problems deter us. I'll illustrate one way to proceed under this worst-case scenario by showing what would happen with a simple theoretical hybrid. I go into the genetics and the whys and wherefores and other breeding approaches and options much more in *Breed Your Own Vegetable Varieties*. For my purposes here, I won't present the calculations. I'll ask you to take my word for the numbers.

Let's consider a simple case where the original cross involves two characteristics, one physical characteristic and one disease-resistance gene. So let's suppose that the original cross was between a pure-breeding determinate variety that is resistant to some theoretical awful disease we care about and an indeterminate variety that is sensitive to the disease. Let's suppose that resistance to the awful disease is conferred by (imaginary) dominant resistance gene *Awf*. And let's suppose the determinate variety is determinate because it carries the (real) recessive gene *sp*, self-pruning. In tomato genetics nomenclature, the ordinary alleles of the gene are designated with a $^+$ superscript. So the original cross to get the hybrid can be represented as *sp sp Awf Awf* × *sp$^+$ sp$^+$ Awf$^+$ Awf$^+$*. And the hybrid is *sp$^+$ sp Awf$^+$ Awf*.

The first step in dehybridizing this hybrid is to simply self-pollinate the hybrid and save the seeds. Tomatoes are usually referred to as self-pollinating, but how much they self versus spontaneously cross varies from variety to variety. So we should plant the hybrid on the other side of the garden from our other tomato plants. Alternatively, if we have several plants of the hybrid, we collect fruits for seed saving from one of the plants in the middle rather than one of those next to plants of another variety. In this first generation it is okay to save seeds from just one plant. So you actually need only one plant of the hybrid for the entire first year of the project.

Saving seed of tomatoes involves cutting them and removing the seeds and juice, or grinding them up briefly in a blender, or smashing them up, then allowing the juice to ferment one to three days to dissolve the gelatin around the seed, destroy germination inhibitors, and eliminate certain diseases. I describe the fermentation process in detail in

Breed Your Own Vegetable Varieties in the tomato genetics part of appendix A (pages 291–96) and so does Ashworth in *Seed to Seed*, so I won't repeat it here. After the fermentation step I wash the seed on a strainer under the kitchen faucet, then prop the strainer up in front of a fan. Then I come by every hour or so and stir and fluff up and rearrange the seed as it dries so that the seeds don't stick to the strainer or to one another.

The second year of the project you plant out the seed you saved (the F_2 seed). I would suggest planting and raising at least eight plants. I would actually prefer sixteen or more, but eight will do. Next collect one or more tomatoes from each plant and pool them for seed saving. Rig things so you will end up with about the same amount of seed from each plant. (In other words, if some plants have tomatoes with less seed add more of those to the pool.) Save seed from the pooled tomatoes. We are taking advantage of the fact that tomatoes are mostly self-pollinating, so all or nearly all the seed will be F_3 seed, but from various parents with various genotypes. If there are a few crosses at this point it won't change things too much.

Raising and collecting seed from more than eight F_2 tomato plants is better than eight. You don't need to give the plants ordinary spacing. You can, for example, crowd sixteen tomato plants into a 4-foot (1.2 m) square of garden space and let each plant set just a couple of fruits, then prune off all other flowers and maybe also all new branches until the fruits mature. The plants will be stunted and the growing conditions would ruin the flavor of the tomatoes, but we aren't evaluating for flavor at this point. (We can't, because much of flavor is associated with recessive genes, and many of those aren't expressing yet at this point in the project.) Now we have F_3 seed.

In years three and four just repeat the growing and seed pooling and saving process. Plant and grow out at least eight plants from the prior generation. Collect and pool fruit from all of the plants each generation. After the four rounds of inbreeding we have F_5 seed. Each generation half the genetic heterozygosity is lost. We end up with an expected F_5 generation of $^{15}/_{32}$ homozygous-resistant + $^{2}/_{32}$ heterozygous-resistant + $^{15}/_{32}$ homozygous-sensitive. The more rounds of inbreeding we perform, the higher the proportion of plants that are homozygous for the resistance gene as well as every other gene involved.

Sooner or later the disease is going to hit our garden and allow us to evaluate the disease resistance of our plants. Let's suppose the disease hits in the year that we are growing out our eight or more F_5 plants. If so, we expect $^{17}/_{32}$ of our plants, that is 53 percent, to be resistant to the disease. But best of all, because of the inbreeding, most of the resistant plants, $^{15}/_{17}$, or 88 percent of them in fact, are now fixed for resistance and will breed true for it. If we planted eight plants, we could expect on average four resistant plants. And on average we'd expect three and a half, that is, three to all four plants, to be pure for resistance.

Now we save seed separately from each of the most resistant plants whose other characteristics we like. Each becomes the foundation plant for a separate line. In the next year we grow eight plants from each line to identify which lines are pure-breeding for resistance as well as to figure out which has the best repertoire of other characteristics we like. At that point we will probably choose just one of the lines to be our new pure-breeding variety.

The same kind of loss of heterogeneity and increase in the proportion of plants that are pure-breeding happens with every other gene in our dehybridizing project including *sp*, self-pruning (that is, determinate growth habit), one of the genes in our theoretical starting hybrid. However, unlike the situation we postulated for the disease resistance, we can actually identify which plants are determinate or indeterminate each generation. If we pay no attention to plant form and randomly save seed from all the plants during the inbreeding phase of the project, we will end up with exactly the same classes and frequencies for the *sp* gene (and all other genes that were involved in the cross) as we did for the disease-resistance gene. After four rounds of inbreeding we would expect $15/32$ homozygous-determinate + $2/32$ heterozygous-indeterminate + $15/32$ homozygous-indeterminate. After four rounds of inbreeding most of the genes involved in the cross will have become fixed, including those for fruit size and color and everything else that might have been involved in the cross that produced the commercial hybrid we started with.

If we want one particular form—indeterminates, for example—we can just choose only that type to save seeds from during the rounds of inbreeding and speed up obtaining purity for that characteristic. Likewise, if the awful disease hit our fields earlier in the project than the F_5 we would have saved seed just from the more resistant tomatoes and speeded things along. If the disease did not hit our fields at the F_5 generation we would continue our cycles of inbreeding and pooling the seeds until the disease did hit our fields and allow us to identify the resistant plants and choose the foundation plant for our new lines.

If we have two or more dominant genes for resistance to a given disease segregating, it changes and complicates the numbers. And the frequencies of the desired classes go down. So we should probably grow sixteen or more plants each generation instead of just eight. But the basic principles and approach to dehybridizing is the same.

In the tomato chapter I listed all the hybrids I could find that are resistant to late blight (*Phytophthora infestans*) and the genetic nature of their resistance where known. There are three major known genes, *Ph1*, *Ph2*, and *Ph3*, but only the latter two are useful against modern late blight lines. All these genes are on different chromosomes. Many of the hybrids are heterozygous for both *Ph2* and *Ph3*. The best resistance so far seems to come from having both those genes. And the genes appear not to be completely dominant. So homozygosity for resistance is better than heterozygosity. If we have the disease in our garden regularly we can speed the project up by breeding from just the most resistant plants each generation. If plants homozygous for resistance are markedly more resistant than those heterozygous for it, we may actually be able to identify homozygous-resistant plants early in the project.

'Hybrid Iron Lady' is especially noteworthy because it is already homozygous for both *Ph2* and *Ph3*. That is, it is already pure-breeding for powerful late blight resistance. All of its offspring can be expected to be late blight resistant. (Reading in between the lines, I think 'Hybrid Iron Lady' is heterozygous for early blight resistance and some other resistances.) So 'Hybrid Iron Lady' should be especially easy to dehybridize into pure-breeding blight-resistant varieties. Since it is already pure-breeding for the blight-resistance genes, we would be dehybridizing to get purity for the other disease resistances and other characteristics.

Many of the blight-resistant hybrids are in commercial varieties that are homozygous for *u*, the flavor-killing gene for uniform ripening, or other genes we don't want. And when a hybrid is homozygous for a gene, we can't get rid of it by inbreeding. We will need to cross to something else first, as described in the final section in this chapter.

Tomato Genes and Genetics

This section is intended to help those who are doing some of their own tomato breeding. Basically, you plug the information here into what you learned in high school genetics (or from *Breed Your Own Vegetable Varieties*) and use it to better design and conduct your tomato breeding projects.

In much of modern genetics genes are named after the way they differ from wild type. With tomatoes, "wild" isn't really the useful concept, so the variety 'Marglobe' is used as the standard. 'Marglobe' is a red-fruited, indeterminate variety. So tomato genes are named after the way they make the plants different from 'Marglobe'. (There are four exceptions I'll list later whose names predate the gene-naming conventions.)

Dominant tomato genes are given symbols that start with uppercase letters. The corresponding ordinary ('Marglobe') allele is named with the same symbol with a $^+$ superscript. Recessive tomato genes are given names that start with lowercase letters. So just seeing the symbol for a tomato gene tells you whether it is dominant or recessive. For example, *sp* is a gene named "self-pruning" that causes determinate plant form. And it was given a symbol starting with a lowercase letter because it is recessive to the ordinary form of the gene (the one found in 'Marglobe'). And

the ordinary allele is designated sp^+. So, when we see the symbol and recall a little of our high school genetics we basically know how this gene will be inherited in crosses. That is, when we cross a determinate variety to an indeterminate one, the cross is $sp\ sp \times sp^+\ sp^+$. The F_1 hybrid is $sp^+\ sp$. And if we self-pollinate the hybrid, the predicted F_2 classes of genotypes and ratios are ¼ $sp^+\ sp^+$ + ½ $sp^+\ sp$ + ¼ $sp\ sp$. However, since the gene is recessive we can't tell the first two classes apart. So the phenotypic (visible) classes we expect in the F_2 are ¾ $sp^+\ 0$ + ¼ $sp\ sp$, that is ¾ indeterminate + ¼ determinate.

Since all these short-vine types involve recessive genes, this means when we cross indeterminate varieties to short-vine types we expect the F_1 to be indeterminate. And since most determinate varieties carry *sp* we would expect most crosses between determinate varieties to give us determinate hybrids. However, if we happened to cross two varieties that are determinate because of different recessive genes we would end up with an indeterminate hybrid.

For a comprehensive list of all known genes in tomatoes and their described effects go to the Tomato Genetics Resource Center on the Internet (http://tgrc.ucdavis.edu) and click on Database Queries, then on Genes, then on List of Gene Names and Symbols.

We often do not know which genes are in a particular variety we are interested in. Many phenotypic characteristics can be caused by any one of a number of different genes. Some genes are common enough so that we can guess the relevant genes involved in the variety, however. Many of these involve tomato color. And sometimes we can get ideas about the way a trait might behave in crosses by just reading through the gene list and seeing if there are known genes that can result in that trait, and, if so, whether they are dominant or recessive.

The List of Gene Names and Symbols also gives the chromosomes the genes are on. If two genes are on different chromosomes they are inherited independently of each other in meiosis and segregate independently in crosses. If they are on the same chromosome they may or may not segregate independently depending on how closely linked they are.

Here are a few of the genes that are often relevant when we do tomato breeding:

Rule Breakers. There are four genes that were discovered and named long before the gene-naming conventions were established and don't follow the rule that genes should be named for how they differ from 'Marglobe'. (They do follow the rest of the rules, including lowercase symbols meaning recessives.) These four genes are *c*, *r*, *s*, and *y*. These genes are associated with potato leaf, yellow flesh, compound inflorescence, and colorless fruit epidermis (skin), respectively.

Determinate Plant Form. The genes commonly involved are *sp*, *br*, and *d*, self-pruning, brachytic, and dwarf, respectively. Most determinate varieties carry *sp*. Some carry *br* or *d* instead of or in addition to *sp*.

Jointless can be associated with *j1* or *j2*. Jointed fruits break off the plant at a particular place on the stem, leaving each tomato with a chunk of stem attached. The fruits of jointless plants separate from the stem at the fruit. (The *br* gene overrides and cancels expression of jointless.)

Fruit Ripening. Uniform ripening is associated with *u*. It removes some of the chlorophyll when the fruit is ripening, eliminates green shoulders on the fruits, cuts down sugar content of the fruit, and apparently wrecks the flavor. It has been incorporated into almost all commercial American hybrids for the sake of the uniform fruit color. Other genes that affect fruit ripening include *nor*, for non-ripening (which is found in 'Longkeeper'), *Nr* for never ripe, *rin* for ripening inhibitor, and *Gr* for green ripe.

Tomato fruit color is affected by the color of the skin (epidermis) and the color of the flesh. Standard red tomatoes have yellow skin and red flesh. Pink tomatoes such as 'Brandywine' or 'Pruden's Purple' have clear skin and red flesh. Yellow tomatoes have yellow skin and yellow flesh. White tomatoes have clear skin and yellow flesh. Clear skin is caused by the *y* gene (one of the naming exceptions). Yellow flesh is caused by *r*, another naming exception. (Yellow-fleshed tomatoes usually have paler-yellow flowers, so you might be able to identify the plants as soon as they start flowering.)

Standard red tomatoes have yellow skin and red flesh so genetically are $y^+ y^+ r^+ r^+$.

Pink tomato varieties have tomatoes with clear skin and red flesh, so are genetically $y y r^+ r^+$.

Yellow tomato varieties have normal yellow skins and yellow flesh, so they are genetically $y^+ y^+ r r$.

White tomato varieties have clear skins and yellow flesh, so they are $y y r r$. (They are actually pale yellow, not white.)

If you remember your high school genetics, you can thus predict what you expect from various crosses. For example, if you cross a variety with white tomatoes to one with red tomatoes and go to the F_2, the cross is $y\,y\,r\,r \times y^+\,y^+\,r^+\,r^+$, which is a two-factor (gene) cross. And consulting the gene list tells you that y is on Chromosome 1 and r is on Chromosome 3. These genes are on different chromosomes so they will segregate independently. Your high school genetics tells you when two factors are simple dominants that segregate independently you get the familiar 9:3:3:1 ratio of classes in the F_2. In this case the F_1 would be y^+ $y\,r^+\,r$. And the F_2 would be $\frac{9}{16}\,y^+$ _ r^+ _ + $\frac{3}{16}\,y^+$ _ $r\,r$ + $\frac{3}{16}\,y\,y\,r^+$ _ + $\frac{1}{16}\,y\,y\,r\,r$. Then looking back at the info on which genes determine which colors you would realize that the F_2 of a cross between red and white tomatoes gives $\frac{9}{16}$ red + $\frac{3}{16}$ yellow + $\frac{3}{16}$ pink + $\frac{1}{16}$ white.

I discuss how to calculate expectations for these (and much more complex crosses) in *Breed Your Own Vegetable Varieties*. To figure out how many plants to grow to be either 95 percent or 99 percent sure of getting at least one of various classes with various probabilities see appendix F (page 340) in *Breed Your Own Vegetable Varieties*.

There are a number of different genes associated with orange fruit color, some of which are dominant and some of which are recessive. Each gene changes the balance between lycopene and various carotenoids in specific ways. You usually aren't going to know which gene you have in any given orange-fruited variety. Of particular interest are B and $Mo\text{-}B$. The B gene is named after beta-carotene content, and is dominant. Plants carrying it have orange fruits with little or no lycopene and five times as much beta-carotene. B is tightly linked to br so does not segregate inde-

pendently with respect to it in crosses. This means when we do a cross involving these two genes we may recover fewer of the recombinant classes than expected. With very tight linkage, as is the case for B and br, we might not get any recombinants at all. So, for example, if we cross an orange determinant variety to a red indeterminate variety, we may get only orange determinants and red indeterminates in the F_2 that is, the parental phenotypes. We may get no orange indeterminates or red determinates.

The $Mo\text{-}B$ gene, modifier of B, that is, increases the amount of beta-carotene even more in the presence of the B gene. We humans can convert beta-carotene to vitamin A, so it is nutritionally significant. The variety 'Caro Red' carries just B. The variety 'Caro Rich' carries both B and $Mo\text{-}B$ and has ten to twelve times as much beta-carotene as red tomatoes.

The t gene, tangerine, gives fruits a rich orange color. It also affects flower color and so can be identified as early as the plant has flowers.

The Del gene causes the fruit to have less lycopene and more delta-carotene.

The gene gf causes the fruit to retain chlorophyll after ripening. When the background color of the tomato is red, this gives you a black, brown, and red mottled fruit. In some variety names such varieties are called black; in others they are called purple. If the background fruit color is yellow, gf gives a tomato that is greenish yellow when ripe. So a black tomato variety is $y^+\,y^+\,r^+\,r^+\,gf\,gf$. And a green-when-ripe tomato variety is genetically $y^+\,y^+\,r\,r\,gf\,gf$.

The gene Gr, that is green-ripe, has effects similar to gf but leaves the center of the fruit unaffected.

Late blight resistance (*Phytophthora infestans* resistance) is conferred by $Ph2$ and $Ph3$. There are dozens of other dominant resistance genes for everything from early blight to nematodes. All the various resistances mentioned on the list of hybrid varieties in the tomato chapter have one or more known dominant genes that have been incorporated into the relevant hybrid. To find these resistance genes, go to the List of Gene Names and Symbols and just scan the left-hand column for symbols starting with capitals until you find the genes relevant to the disease you care about.

The genes I mentioned here and the genes on the master list at the Tomato Genetic Resource Center are not all the genes there are. You might find yourself working with other genes that segregate in your crosses. Furthermore, many of the characteristics we care about the most are affected by many genes, each of which may have only a small effect. When this is the case, when we do crosses we get variability for the characteristic in the offspring, but don't see clean segregation classes. Plant growth rate, vigor, earliness, fruit size, and flavor, for example, are affected by many genes.

The same basic rules apply to working with unknown genes or quantitative characters as with the familiar genes with big, discrete effects. Basically we use our tools of doing crosses, inbreeding, backcrossing, and selecting whatever the nature of the genetic variability we are manipulating.

Breeding the Heirloom Tomatoes of Tomorrow

Dehybridizing hybrid blight-resistant tomatoes to create new varieties is most practical if you like and use those hybrid varieties. However, many of us prefer the spectacular flavors and all the colors and shapes of heirloom tomatoes. And we don't mind that they are irregularly shaped and colored and are usually too soft to transport well. We don't need to transport them any farther than from our garden to our table. However, in the future we are probably going to be unable to grow most or all of our current repertoire of heirloom varieties because of late blight. (See the tomato chapter.) So how do we breed tomatoes that have the spectacular flavors of our favorite heirlooms but that are also resistant to late blight?

One basic approach to creating disease-resistant heirloom-style varieties is to look for and take advantage of any disease resistance that may already exist in heirloom and other open-pollinated varieties. It is clear that most heirlooms are highly susceptible to late blight. However, a few heirloom or open-pollinated varieties have been reported as having significant resistance. I compiled and presented a list of these in the tomato chapter. There may be additional heirloom or open-pollinated varieties you can discover yourself just by trying a few new varieties every year. Once you know of two varieties you like that show at least some resistance to late blight under your conditions, cross them, grow out the hybrid, save seed of the hybrid to get F_2 seed, grow the F_2 plants, and start selecting for disease resistance. The object is to select one or more new varieties that are more resistant than either parent. This approach has the advantage that it might be able to produce varieties with "horizontal" as opposed to "vertical" resistance.

So-called vertical resistance is associated with single genes, usually dominant genes transferred by familial breeding. Where resistance is based on a single gene it tends to be easier for the pest to overcome. The term *horizontal resistance* assumes many genes contributing to resistance. The variety may not be completely resistant but is resistant enough. Fans of heirlooms and traditional varieties often tout their disease resistance (when they have it) as being more "durable" than that associated with single genes. Actually, I think the categories are imaginary at worst and overlap a good bit at best. We simply don't know the genetic basis of the resistance in heirloom crop varieties. And by the time we incorporate two dominant resistance genes into a variety by familial breeding, the resistance is at that point multigenic, and should be harder for the pest to overcome, so is likely to be more durable. So is the resistance conferred by two dominant genes vertical or horizontal?

Whatever reality there is or isn't about the distinction between vertical and horizontal resistance, however, if we cross two resistant heirlooms, go to the F_2, then start selecting for resistance, we can select for recessive genes and multigenic characteristics that we lose when we focus just on transferring dominant genes. Best of all, when we cross two heirlooms that taste great, we are pretty likely to be able to select out new varieties that also taste great.

To learn how to make tomato crosses, go to the Tomato Genetics Resource Center on the Internet (http://tgrc.ucdavis.edu) and click on Stock Maintenance Guidelines, then click on the article "Guidelines for Emasculating and Pollinating Tomato Flowers" by Roger Chetelat and Scott Peacock. (I also cover how to make tomato crosses in *Breed Your Own Vegetable Varieties* in appendix A, pages 291–96, with an illustration on page 306. However, the aforementioned article goes into more detail and has an excellent series of full-color photos.)

The tomato flower has a style (female part) that is surrounded by a cylinder ("cone") of fused anthers that have their pollen-bearing surfaces facing inward. To do a cross we need to remove that anther cone before it releases any pollen, then transfer pollen of the type we want to the stigma (the receptive part of the female flower, which is located at the top of the style) of our emasculated flower as illustrated in the article. The best time for removing the anther cone is at the late bud stage when the bud is just beginning to turn yellow but has not opened yet. The best time to collect pollen from a flower to use in making our own pollinations is right after the flower opens. For details see the article (or the information in *Breed Your Own Vegetable Varieties*).

Tomato crosses are more likely to succeed if they are made on one or two of the biggest flowers in a cyme rather than on the smaller flowers. You should remove all the uncrossed flowers from the cyme so that they do not compete with the crossed flowers. With other crops I've found hand pollinations succeed at a much higher rate when they are made early in the flowering season on the earlier, bigger flowers, and when they have no competition at all with preexisting developing fruits. I don't know that this is true for tomatoes, but I'm guessing it probably is. So I would suggest removing all earlier fruits so that your cross will be the very first fruit on the plant. Crossed flowers don't have to be bagged to protect them from insects because insects are not attracted to the flowers once the pollen-bearing anther cylinder and petals are removed. You'll need to tag the cyme with the crosses in some way so you can identify the crossed tomatoes later.

A third approach to creating disease-resistant heirloom-style varieties is to cross our favorite heirloom to a commercial hybrid that contains one or more dominant genes for resistance, go to the F_2, then start selecting in each generation for disease resistance and the colors and flavors we want. (Since most hybrids are heterozygous for disease-resistance genes, we have to choose one or more resistant F_1 plants to go on with. Not all the F_1 plants will be resistant.)

However, it takes lots of different genes to get the flavors we want, and in my experience breeding for flavor, many of the flavor-associated genes are recessive. In addition the commercial hybrids introduce lots of genes associated with inferior flavor. The more genes you have to have in specific combinations, the rarer the class among the offspring. So it can be hard to select out material with good flavor after crossing a variety with great flavor to something with poor flavor. A variant approach I often use to select varieties with superb flavor is to do a backcross to the delicious parent. That is, instead of making the hybrid and self-pollinating it (or allowing it to self) to get the F_2, I backcross the F_1 plant to the delicious parent. Then I start selecting by choosing foundation plants and inbreeding from there. The advantage to the backcross is that after the backcross, the resulting seed has three-fourths of its genes from the delicious parent instead of only half. In addition, all the recessive characteristics are more likely to show up and are more frequent. (Each characteristic associated with a recessive gene appears half of the time after the backcross instead of one-fourth of the time as it does in the F_2.) This makes it much easier to select out delicious stuff in subsequent generations.

A final approach to creating new late blight resistant heirloom-style varieties involves transferring the dominant resistance genes from a commercial hybrid or other variety into our favorite heirloom, deliberately throwing away most or all of the rest of the genes in the commercial variety. This approach makes a lot of sense when you love the heirloom variety and hate everything about the hybrid except for its disease resistance. To transfer dominant genes from one variety into another using traditional plant breeding methods we use an approach called recurrent backcrossing.

To do recurrent backcrossing we start by crossing the variety most of whose genes we want to keep (the heirloom in this case) to the variety that is a source of the dominant genes we want to transfer. Then we backcross the F_1 to the heirloom. Then we choose the most disease-resistant plant among the offspring from that cross and backcross it to the heirloom again. And so on. Each time we backcross to the heirloom we are discarding half the genes from the other variety except for genes we specifically select for. We specifically select for disease resistance each generation, so retain those genes. After about six rounds of backcrossing, our variety has almost all its genes from the heirloom except for the dominant genes we specifically select for each generation (the disease-resistant genes in this case). The exception is likely to be genes that are very close to the selected dominant genes on the same chromosome; these are likely to be transferred as a block along with the selected dominant genes. So we always end up with a variety that is at least a little different from the recurrent parent. After six rounds of backcrossing, we will end up with plants that are nearly identical to the heirloom, even if we do no particular selecting for the characteristics of the heirloom along the way. The only selecting we need to do each generation is for the disease resistance we are transferring into the heirloom.

Since there are two known effective dominant genes for late blight resistance (*Ph2* and *Ph3*) and the best resistance involves multiple genes, I suggest starting by crossing your favorite heirloom to a variety that contains both *Ph2* and *Ph3*. At this point, the only released varieties that contain both genes are commercial hybrids. I list the commercial hybrids and the genes they contain in the tomato chapter. Using 'Iron Lady Hybrid' as the source of dominant resistance genes is particularly advantageous, however, because it is homozygous for both genes. This means all offspring of the initial cross between your heirloom and the hybrid will be heterozygous for both genes. If instead you start by crossing to a hybrid that is heterozygous for both genes, only one-fourth of the offspring would be heterozygous for both. So you would have to grow multiple plants that first generation and evaluate them for resistance instead of being confident that all plants carried both resistances.

After you do your initial cross, save seeds from the crossed fruit and plant them the following year. In a cross between an heirloom that doesn't carry either dominant resistance gene and 'Iron Lady' we can represent the cross as $Ph2^+ Ph2^+ Ph3^+ Ph3^+ \times Ph2\ Ph2\ Ph3\ Ph3$. The F_1 hybrid is heterozygous for both resistance genes, and is $Ph2^+ Ph2\ Ph3^+ Ph3$. Now we choose the most disease-resistant plant among the offspring and backcross it to the heirloom again and save the seeds. The backcross can be represented as $Ph2^+ Ph2\ Ph3^+ Ph3 \times Ph2^+ Ph2^+ Ph3^+ Ph3^+$. The expected classes among the offspring are one-fourth each of all possible classes, with the class that carries both genes for resistance being the most valuable. So we grow enough plants so we are likely to get at least one of the double-resistant genotype to continue breeding from. How many is that? I look this up in appendix F in *Breed Your Own Vegetable Varieties* and find that if I grow eleven plants I can be 95 percent sure of getting at least one of the class where the probability of the class is .25. So grow at least eleven plants, choose the one that is the most disease-resistant, and backcross it to the heirloom again. Each generation grow out at least eleven plants, and continue the project with the plant among them that appears to be the most disease-resistant.

Growing eleven plants need not take much space. You can evaluate disease resistance on plants that are too crowded to produce good fruits. You can't evaluate flavor on plants so grown. However, during the backcrossing phase, you don't need to evaluate flavor. You are simply trying to generate flavor that is as close to the heirloom as possible, and are depending on the statistics of the backcross process to achieve that. This means that during each generation of the breeding project you can jam lots of tomato plants in a small bed and let them grow just big enough to display their resistance and to produce flowers from which to obtain pollen. You can use that pollen on properly grown and spaced plants of the heirloom that is the recurrent parent, which, being your favorite, you are probably growing for food anyway. You can grow sixteen crowded breeding-project plants in a 4-foot (1.2 m) square of garden space. So the project of transferring dominant resistance genes into your favorite heirloom need take only that amount of space each year above the ordinary amount of space you spend on growing the heirloom itself for food.

You may continue the backcrossing and selecting for six generations to get material that is nearly identical for all the genes in the heirloom other than the selected resistance genes and genes closely linked to them. By the fourth generation of backcrossing you should have material that looks very similar to your heirloom and that probably includes many plants that taste quite similar also but is still a bit heterogeneous. You may want to stop backcrossing at that point and start saving seeds from individual plants selected for everything you care about, including flavor. You will need at least a couple of rounds of inbreeding to get the disease-resistance genes from heterozygous, as they are during the backcrossing phase, to homozygous, as they need to be in your new pure-breeding variety.

Once you have your new variety, save plenty of seed. Rejoice in it. Cherish it. Share it. Pass it along.

Congratulations. You have bred a variety that may become one of the heirloom tomatoes of tomorrow.

Seed Companies and Sources

Here are the seed companies mentioned in this book, plus a few others. This list includes many small one-person or one-family operations that are among the prime breeders and preservers of open-pollinated public domain varieties. These companies often have website-only catalogs and take orders only via mail order or website, depending on the company. Where I list no phone number it is because the company is not set up for and does not take phone orders. This list includes U.S. and Canadian seed sources only. Canadian companies have CANADA immediately after the company name. Some U.S. companies ship to Canada and some Canadian companies ship to the United States. Where a person associated with the company is mentioned in or provided information for this book, he or she is listed in parentheses.

Adaptive Seeds—Seeds of the Seed Ambassadors Project (Andrew Still and Sarah Kleeger). www.adaptiveseeds.com. 25079 Brush Creek Road, Sweet Home, OR 97386.

Baker Creek Heirloom Seeds. www.rareseeds.com. 417-924-8917. 2278 Baker Creek Road, Mansfield, MO 65704.

Bountiful Gardens. (John Jeavons of Bountiful Gardens is author of *How to Grow More Vegetables (and Fruits, Nuts, Berries, Grains, and Other Crops) than You Ever Thought Possible on Less Land than You Can Imagine*, 8th edition, and co-author of *The Sustainable Vegetable Garden: A Backyard Guide to Healthy Soil and Higher Yields*.) www.bountifulgardens.org. 707-459-6410. 18001 Shafer Ranch Road, Willits, CA 95490.

Fedco Seeds. (CR Lawn of Fedco provided information about especially flavorful tomatoes for New England.) www.fedcoseeds.com. P.O. Box 520, Waterville, ME 04903.

Fertile Valley Seeds (Carol Deppe, author of *Breed Your Own Vegetable Varieties: The Gardener's and Farmer's Guide to Plant Breeding and Seed Saving*, *The Resilient Gardener: Food Production and Self-Reliance in Uncertain Times*, and *The Tao of Vegetable Gardening*). www.caroldeppe.com or www.fertilevalleyseeds.com. No phone orders. To receive a catalog and ordering information by email send your request to carol@resilientgardener.com. 7263 Northwest Valley View Drive, Corvallis, OR 97330.

High Mowing Organic Seeds. www.highmowingseeds.com. 802-472-6174. 76 Quarry Road, Wolcott, VT 05680.

Johnny's Selected Seeds. (Rob Johnston. I referred to the Johnny's catalog for information about optimal tomato seed germination temperatures, dealing with late blight, and much else. Very detailed growing information.) www.johnnyseeds.com. 1-877-564-6697. 955 Benton Avenue, Winslow, ME 04901.

Joseph's Garden (Joseph Lofthouse). Landraces that Joseph has bred for his Utah mountain farm (including the 'Lofthouse Landrace Moschata' I discussed in the squash chapter). Garden@Lofthouse.com. http://garden .lofthouse.com. Joseph Lofthouse, P.O. Box 538, Paradise, UT 84328.

Native Seeds/SEARCH. www.nativeseeds.org. 520-622-5561. 3061 North Campbell Avenue, Tucson, AZ 85719.

Nichols Garden Nursery. (Rose Marie Nichols McGee and Keane McGee. Rose Marie is co-author of *The Bountiful Container: Create Container Gardens of Vegetables, Herbs, Fruits, and Edible Flowers*.) www.nicholsgardennursery .com. 1-800-422-3985. 1190 Old Salem Road Northeast, Albany, OR 97321-4580.

Peace Seedlings (Dylana Kapuler and Mario DiBenedetto). Also see Peace Seeds. Note: The two Kapuler family seed companies have lists that overlap, but each has plenty that is unique. www.peaceseedlings.com. 2385 Southeast Thompson Street, Corvallis, OR 97333.

Peace Seeds (Alan Kapuler). Also see Peace Seedlings. Note: The two Kapuler family seed companies have lists that overlap, but each has plenty that is unique. www.peaceseeds.com. 2385 Southeast Thompson Street, Corvallis, OR 97333.

Peaceful Valley Farm Supply, Inc. www.grow organic.com. 1-888-784-1722. P.O. Box 2209, Grass Valley, CA 95945.

Pinetree Garden Seeds. www.superseeds.com. 207-926-3400. P.O. Box 300, New Gloucester, ME 04260.

Redwood City Seed Company. www.ecoseeds .com. P.O. Box 361, Redwood City, CA 94064.

Renee's Garden Seeds. www.reneesgarden.com. 6060A Graham Hill Road, Felton, CA 95018.

Salt Spring Seeds. CANADA. www.saltspringseeds .com. Box 444, Ganges P.O., Salt Spring Island, BC, V8K 2W1, Canada.

Sand Hill Preservation Center—Heirloom Seeds & Poultry. www.sandhillpreservation.com. 1878 230th Street, Calamus, IA 52729.

Seed Ambassadors Project. www.seedambassadors .org. 25079 Brush Creek Road, Sweet Home, OR 97386.

Seeds of Change. www.seedsofchange.com. 1-888-762-7333. P.O. Box 4908, Rancho Dominguez, CA 90220.

Seeds of Diversity. CANADA. Joining this Canadian seed savers exchange gives you access to more than two thousand open-pollinated varieties maintained by its members. www .seeds.ca. 905-372-8983. P.O. Box 36, Station Q, Toronto, ON, M4T 2L7, Canada.

Seed Dreams. gowantoseed@yahoo.com. P.O. Box 106, Port Townsend, WA 98368.

Seed Savers Exchange. Both a seed company and an exchange. Getting on the free mailing list only gets you the seed company offerings. You have to join in order to get access to the thousands of varieties grown by members. Many small seed companies distribute exclusively via listings in the SSE Winter Yearbook (which comes with membership). www.seedsavers .org. 563-382-5990. 3094 North Winn Road, Decorah, IA 52101.

Southern Exposure Seed Exchange. (Ira Wallace of SESE is author of *The Timber Press Guide to Vegetable Gardening in the Southeast*.) www.southernexposure.com. 540-894-9480. P.O. Box 460, Mineral, VA 23117.

Tater Mater Seeds (Tom Wagner and Rob Wagner). This company represents the tomato and potato breeding work of Tom Wagner. A source of many heirloom-quality but new tomato varieties, as well as true potato seed. www.tatermaterseeds.com. 8407 18th Avenue West, 7–203, Everett, WA 98204.

Tatiana's Tomatobase—Heritage Tomatoes. CANADA. This is a collection of more than four thousand open-pollinated tomatoes with sources, breeders, country of origin, descriptions, and photos. The collection is funded by the Tatiana's Tomatobase Seed Shop, which each year offers for sale a subset of the tomatoes in the collection. Website catalog and orders only. www.tatianastomatobase.com. Anmore, BC, Canada. (No more exact address is available.)

Territorial Seed Company. (Steve Solomon, author of *Growing Vegetables West of the Cascades*, *Gardening When It Counts*, and *The Intelligent Gardener*, founded Territorial.) www.territorialseed.com. 541-942-9547. P.O. Box 158, Cottage Grove, OR 97424-0061.

Totally Tomatoes. www.totallytomato.com. P.O. Box 295, 334 West Stroud Street, Randolph, WI 53956.

Trade Winds Fruit. Huge list of tomato varieties. www.tradewindsfruit.com. P.O. Box 9396, Santa Rosa, CA 95405.

Uprising Seeds. www.uprisingorganics.com. 2208 Iron Street, Bellingham, WA 98225.

Victory Seeds. www.victoryseeds.com. P.O. Box 192, Molalla, OR 97038.

West Coast Seeds. CANADA. www.westcoastseeds .com. 1-888-804-8820. 3925 64th Street, RR#1, Delta, BC, V4K 3N2, Canada.

Wild Garden Seeds (Frank Morton). www.wild gardenseed.com. P.O. Box 1509, Philomath, OR 97370.

William Dam Seeds. CANADA. www.damseeds .com. 905-628-6641. 279 Highway 8, Dundas ON, L9H 5E1, Canada.

Index

Note: *ci* page references can be found in the color insert.

ABOUT THE AUTHOR

Keane McGee / Nichols Garden Nursery

Oregon plant breeder Carol Deppe holds a PhD in biology from Harvard University and specializes in developing public-domain crops for organic growing conditions, sustainable agriculture, and human survival for the next thousand years. Carol is author of *The Resilient Gardener: Food Production and Self-Reliance in Uncertain Times* (Chelsea Green, 2010), *Breed Your Own Vegetable Varieties: The Gardener's and Farmer's Guide to Plant Breeding and Seed Saving* (Chelsea Green, 2000), *Tao Te Ching: A Window to the Tao through the Words of Lao Tzu* (Fertile Valley Publishing, 2010), and *Taoist Stories: A Window to the Tao through the Tales of Chuang Tzu and Lieh Tzu* (Fertile Valley Publishing, 2013). Visit www .caroldeppe.com to buy seeds and for articles and further adventures.

green press
INITIATIVE

Chelsea Green Publishing is committed to preserving ancient forests and natural resources. We elected to print this title on 100-percent postconsumer recycled paper, processed chlorine-free. As a result, for this printing, we have saved:

125 Trees (40' tall and 6-8" diameter)
56 Million BTUs of Total Energy
10,766 Pounds of Greenhouse Gases
58,392 Gallons of Wastewater
3,909 Pounds of Solid Waste

Chelsea Green Publishing made this paper choice because we and our printer, Thomson-Shore, Inc., are members of the Green Press Initiative, a nonprofit program dedicated to supporting authors, publishers, and suppliers in their efforts to reduce their use of fiber obtained from endangered forests. For more information, visit: www.greenpressinitiative.org.

Environmental impact estimates were made using the Environmental Defense Paper Calculator. For more information visit: www.papercalculator.org.

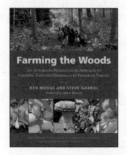

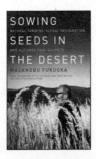

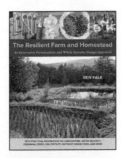